Edsel Ford and E.T. Gregorie

The Remarkable Design Team
and Their Classic Fords of the
1930s and 1940s

Other related SAE books of interest:

The Birth of Chrysler Corporation and Its Engineering Legacy
by Carl Breer
(Order No. R-144)

The Franklin Automobile Company
by Sinclair Powell
(Order No. R-208)

Automobile Design: Twelve Great Designers and Their Work
Edited by Ronald Barker and Anthony Harding
(Order No. R-115)

For more information or to order this book, contact SAE at 400 Commonwealth Drive, Warrendale, PA 15096-0001; (724) 776-4970; fax (724) 776-0790; e-mail: publications@sae.org.

Edsel Ford and E.T. Gregorie

The Remarkable Design Team and Their Classic Fords of the 1930s and 1940s

HENRY DOMINGUEZ

Society of Automotive Engineers, Inc.
Warrendale, Pa.

Copyright © 1999 Society of Automotive Engineers, Inc.

Library of Congress Cataloging-in-Publication Data

Dominguez, Henry L., 1953-
    Edsel Ford and E.T. Gregorie: The remarkable design team and their classic Fords of the 1930s and 1940s / Henry Dominguez.
    p. cm.
    Includes bibliographical references and index.
    ISBN 0-7680-0400-4
    1. Ford automobile--Design and construction--History. 2. Ford, Edsel, 1893-1943. 3. Gregorie, E.T. I. Title.

TL215.F7D65 1999   99-13062
629.222'2--dc21   CIP

Copyright © 1999   Society of Automotive Engineers, Inc.
                       400 Commonwealth Drive
                       Warrendale, PA 15096-0001 U.S.A.
                       Phone: (724) 776-4841
                       Fax: (724) 776-5760
                       E-mail: publications@sae.org
                       http://www.sae.org

ISBN 0-7680-0400-4

All rights reserved. Printed in the United States of America.

Permission to photocopy for internal or personal use, or the internal or personal use of specific clients, is granted by SAE for libraries and other users registered with the Copyright Clearance Center (CCC), provided that the base fee of $.50 per page is paid directly to CCC, 222 Rosewood Dr., Danvers, MA 01923. Special requests should be addressed to the SAE Publications Group. 0-7680-0400-4/99-$.50.

SAE Order No. R-245

*To my wife, Pat DeHerrera,
and my son Samuel*

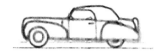

ACKNOWLEDGMENTS

This book has been made possible only because of the kindness and candor of Mr. E.T. Gregorie and the untold hours he gave to me in recounting his design career at Ford Motor Company. For that, I will be forever grateful.

I also want to thank the vivacious Mrs. Evelyn Gregorie for her hospitality while I visited her comfortable and inviting home. She fed me, gave me drink, and endured the Ford stories, all of which she has probably heard a thousand times.

This book would have been possible, but not nearly as informative and lively, were it not for the open and honest conversations I had with Mr. Gregorie's "boys": John Najjar, Emmett O'Rear, Tucker Madawick, Bob Thomas, John Hay, Benny Barbera, Ross Cousins, Wayne Booth, and Andrew Olinik. Likewise, my conversations with former Briggs designer, Rhys Miller, and the legendary Ralph Roberts were simply precious.

Back in 1978, I first met David Crippen, who was then Chief Archivist at the Ford Archives. I had an idea for a Ford book, and he encouraged me to write it. *The Ford Agency* was the result, and I completed that book only because of Dave's unflagging support. In subsequent years, we have become good friends, and Dave helped me tremendously in conducting research for another Ford book, which is under development, but from which this book was conceived.

All other members of the Ford Archives staff have also been extremely supportive and helpful. First, there is Liz Jordan (I call her Betty). She is a warm and wonderful person who travels the world with her daughter. Betty has been at the Ford Archives for almost 30 years. If anybody could write the definitive Ford biography, it would be Betty. Cynthia Read-Miller is Curator of the massive photograph collection at the Ford Archives. I cannot begin to thank her enough for leading me to many of the marvelous photographs in this book. In addition, she made it possible for me to visit other photographic archives in the Detroit area. Thanks also to Steven Hamp, Charles Hanson, Linda Skolarus, Alene Soloway, Denise Moline, and Cathleen Lantendresse for their support and hard work in providing materials from the vast collection of the Ford Archives. To John Tobin and Luke Gillium-Swetman, who are no longer at the Ford Archives but helped me when they were there, I extend my gratitude.

My appreciation also is extended to William Clay Ford, Jr., Chairman, Ford Motor Company; Charles Snearly, Ford Public Affairs Office; Darlene Flaherty, Ford Corporate Archivist; and Dan Erikson, Ford Photomedia Department. Thanks also to Pat Zacharias of *The Detroit News*.

Doctor Paul Weisser was my editor. He and I spent countless hours at his computer, going over the text of this book. By the time we had gone through the manuscript a few times, Paul had learned much about cars and car design, and I had learned much about Webster's inconsistencies.

When Edsel Ford's wife, Eleanor, died in 1976, she left her lovely Cotswold mansion at Gaukler Pointe for posterity. Within this lovely home are most of the possessions Eleanor and Edsel had collected over the years, including some marvelous photographs, which are maintained by Maureen Devine. I want to thank Maureen and also James Bridenstine for allowing me to peruse the archival material at the Edsel & Eleanor Ford House, and Donna Buchanan for being such a gracious host.

I have been studying Ford history for more than 20 years, but I am always amazed at the knowledge of my good friends Dick Folsom, Mike Skinner, and Ford Bryant. Dick has had the privilege of interviewing a number of Henry Ford's relatives and others who knew Henry, Clara, and Edsel, and he graciously put me in contact with many of them. In his eclectic collection of Ford memorabilia, Dick has the only known, privately owned document that has both Henry Ford's and Clara Ford's signatures on it. Mike Skinner also is an avid "Fordophile," and he maintains an extensive Ford memorabilia collection. Between Dick and Mike, they have enough Ford memorabilia to establish their own Ford archives. Ford Bryant's marvelous books on the Ford family and Henry Ford's interests fill in many holes left by other Ford biographers.

Finally, I offer my heartfelt thanks to Mike Lamm, an automotive writer par excellence. With only one phone call, Mike convinced Mr. Gregorie to talk to me. I had tried for five years to obtain Mr. Gregorie's consent for an interview but to no avail. However, Mike has known Mr. Gregorie for many years and put me in touch with him in only five minutes. Without Mike's phone call, this book would not have been possible.

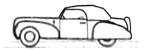

CONTENTS

Foreword xiii

Preface xv

Introduction xvii

Chapter 1   Mr. Gregorie and Mr. Ford
                 Designer and Patron 1

Chapter 2   Ford Hires a Designer
                 "I've Been in Body Design at Brewster" 11

Chapter 3   Edsel Ford
                 Styling Executive 21

Chapter 4   The Ford Model Y
                 Gregorie's First Ford Design 57

Chapter 5   The 1933 and 1934 Fords
                 "Like a Man's Face" 63

Chapter 6	The Continental Car I *"Long, Low, and Rakish"*	67
Chapter 7	The Continental Car II *Lower and Longer*	71
Chapter 8	The 1936 Lincoln-Zephyr *"The First Successfully Streamlined Car in America"*	75
Chapter 9	The Continental Car III *"I Like Right-Hand-Drive Cars"*	81
Chapter 10	The Ford Design Department *"Anything the Boy Wants"*	91
Chapter 11	The 1935–1937 Fords *Years of Transition*	139
Chapter 12	The 1938 Lincoln-Zephyr *"It's Going to Ruin Us"*	149
Chapter 13	The 1939 Lincoln-Zephyr Continental *"The Most Beautiful Car in the World"*	155
Chapter 14	The 1938–1940 Fords *"From an Old Car to a New Car"*	169
Chapter 15	The 1939 Mercury *"Just a Stylized Ford"*	201
Chapter 16	The 1941–1948 Fords and Mercurys *"That Hangdog Look"*	213
Chapter 17	Postwar Development *"A Little Ford and a Big Ford"*	247

Chapter 18 The 1949 Mercury and Lincoln
The Last Hurrah 283

Chapter 19 The End of an Era
"...An Extended Vacation" 309

Bibliography 319

Index 323

About the Author 333

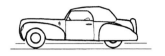

FOREWORD

My grandfather, Edsel Ford, once told a friend, "My father made the most popular car in the world. I'd like to build the best car in the world." To help him build that car—the Lincoln Continental—Edsel called on the talents of E.T. "Bob" Gregorie, the first design chief at Ford Motor Company.

This book describes the creative collaboration that produced the Continental and many other classic cars of the 1930s and 1940s. As head of the newly formed Design Department at Ford, Bob Gregorie had a hand in the design of nearly every Ford, Mercury, Zephyr, and Lincoln built in that period. Both he and my grandfather were gifted with an artistic vision and a genius for automotive styling. Together they created some of the most innovative and beautiful designs in automotive history.

This book brings to life the struggles and triumphs of these two automotive pioneers, and their unique and highly successful working relationship. It details a fascinating era in automotive history and the people who made that history.

William Clay Ford, Jr.
Chairman, Ford Motor Company

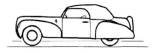

PREFACE

More than half a century has passed since I left Ford Motor Company, and as the company's first design chief, I have received many requests from automotive writers over those years to recount my experiences at Ford. However, until now, I have been reluctant, partly because it was such a small part of my life experience and partly because I did not want to glorify something that, to me, was simply a job.

This book takes a different tack than all other books on the design aspect of Ford products. It emphasizes the working relationship between Mr. Edsel Ford and me, and how, together, we designed automobiles at Ford Motor Company. Although I did the design work—taking pencil in hand and sketching cars—Edsel Ford's innate design ability steered me toward those wonderful old Ford, Mercury, and Lincoln designs.

As I reminisce about my time at Ford, I still cannot explain why Edsel Ford's contributions to Ford design have been largely overlooked. All that I have ever read about Edsel's involvement in the Ford Motor Company ignores his contributions and accentuates his inability to stand up to his father. Although I cannot comment on other aspects of the business (because I was not privy to those discussions), I can tell you from personal experience that, as far as design and product development were concerned, Edsel Ford took a Herculean stand against his father and won major victories in these areas. In fact, it was only because Edsel had the fortitude to stand up to his father that he was able to establish his own design department within the company, allowing us to introduce the Lincoln-Zephyr, the Mercury, and the Continental and to develop numerous facelifts. Edsel Ford pushed through all these products at great personal sacrifice. That fact should never be forgotten. If it weren't for Edsel Ford, I hate to think what would have happened to the Ford Motor Company.

Edsel Ford has been the only president of an automobile company who took such an in-depth interest in design. For example, Alfred Sloan at General Motors realized the importance of automobile design; however, after he hired Harley Earl to direct the company's Art & Colour Section, Sloan disassociated himself from the design process and let Earl do all the creating. Since then, all automobile companies have set up their design departments in this way—except when I worked at Ford Motor Company. When Edsel Ford made me design chief, he did not abdicate his interest in design. He merely assumed the added responsibility of design director, along with his burdensome responsibility as president. This arrangement—one that had never existed previously and will most certainly never exist again—made our working relationship unique. With Edsel's full command of Ford Motor Company and its unlimited resources, we were able to sit down together and develop automobile designs that were to our liking. After we hashed out a design, all Edsel Ford had to do was give the okay, and the production process began. It was that simple.

Through Edsel's dual role as president and design director, we were able to place the Lincoln Continental into production in less than 12 months after I first showed Edsel the rudimentary sketch that I had drawn one morning. Even in those days, such speed was unheard of; today, getting a new car into production that quickly would be impossible. However, the Continental is only one example of the many designs on which Mr. Edsel Ford and I worked over the years—designs that were done quickly, efficiently, and without fanfare. That is the way he liked to work, and that is the way I liked to work. We made a good team.

I am glad that Edsel Ford's significant contributions are now recorded for posterity and that I was able to help tell his story. It was a pleasure to have worked for him, and any accolades that I received over the years as Ford Motor Company's first design chief I must attribute to Mr. Edsel Ford.

E.T. "Bob" Gregorie

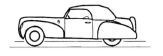

INTRODUCTION

It's a mystery why one person develops an interest in automobiles, another develops an interest in art, and another develops an interest in architecture. For me, my childhood fascination with trains quickly and abruptly changed to automobiles, specifically Ford automobiles, a year or two before I reached driving age. My first car was a 1946 Ford business coupe, and I wanted a Ford automobile not because it was inexpensive to buy—which this coupe was—but because of its styling. In my mind, the styling of Ford automobiles, especially those built between 1932 and 1948, had the simplicity of line and fineness of form that no other automobiles of that era had. Automobiles made by General Motors and Chrysler might have had more refined engineering; however, in terms of styling, none of their products were more beautiful than the Fords, Mercurys, and Lincolns of the 1930s and 1940s.

As I grew older, my interest in Ford automobiles expanded into the history of the Ford Motor Company. After reading every book on Ford history that I could find, I realized that two major aspects of the company had not been explored to any great extent: (1) the role of the Ford dealership organization in the development and success of the company, and (2) why Ford automobiles of the 1930s and 1940s were designed so beautifully.

I addressed the first void with my book *The Ford Agency*, and several years ago, I began researching the design history of those good old Fords. Coincidentally and fortunately, the Ford Archives had already begun tracking down some of the old-time designers, many of whom were well into their eighties, and had conducted interviews with them through the auspices of the Edsel B. Ford Design History Center. Dave Crippen, former Curator of Design History at the Ford Archives, had assisted in my research for the dealer book and had conducted these interviews. He told me that the two men responsible for the design of most early Ford products were Edsel Ford and E.T. Gregorie. Edsel Ford had passed away long ago, but Dave informed me that Gregorie was alive and well, and suggested that I get in touch with him.

Getting in touch with the great designer was easy. Getting him to talk to me was another matter. As I soon discovered, Mr. Gregorie is a man who likes his privacy. In fact, I wrote letters to him for years before he consented to talk with me. However, it wasn't my persistence that won me an interview with Mr. Gregorie; rather, it resulted from a phone call to him on my behalf from my good friend and noted automotive writer Michael Lamm. After my five years of fruitless endeavors, Mike called Mr. Gregorie, and twenty minutes later, I was talking to the great designer.

Initially, our conversations centered around Mr. Gregorie's knowledge of Edsel Ford, but those conversations inevitably led to how he and Edsel worked together to develop designs for Ford products. We spent a lot of time talking about how Mr. Gregorie created his most famous design—the fabulous Lincoln Continental. It was surprising how his account—which, of course, is the true account—differed so dramatically from conventional literature. With the exception of a few obscure articles, the stories of how the Lincoln Continental was developed, as described in popular and respected automobile literature, were absolutely untrue! It was also surprising that the popular automotive literature had neglected to explain how the other fabulous early Fords, Mercurys, and Lincolns were designed. The information these sources proffered centered primarily on mechanical, production, and performance characteristics. No in-depth analysis had ever been done regarding how these early Ford products were designed. By the time I realized this, I had established a good rapport with Mr. Gregorie and decided to review with him every major design of his career, recording on paper the how and why of every Ford product designed under his tutelage.

The first step in this process was to obtain detailed photographs of all of Gregorie's designs from the Ford Archives at the Henry Ford Museum & Greenfield Village in Dearborn, Michigan. The Ford Archives is to the "Fordophile" what Falling Water is to an architect or the Louvre is to an art historian. The story goes that, after the death of Henry Ford's wife, Clara, in 1950, a team of historians went to the 56-room mansion to see what old Henry might have saved over the years. To their surprise, he saved everything! Room after room was crammed full of material—from hand tools, clocks, and rare books to boxes filled with papers, letters, and photographs. There were thousands and thousands of photographs—photographs of Ford's family dating back to the 1860s, and photographs of Ford automobiles covering the entire development of the company. This material eventually was organized into what is now known as the Ford Archives. Added to it were collections from dusty files of the Ford Motor Company, which now has an interest toward the future, not the past. The result is a collection of automobile and industrial history unequaled anywhere in the world.

The staff of the Ford Archives has always been gracious and supportive of my research endeavors since I first stepped into their hallowed halls more than two decades ago. When I approached them with the idea of this project, they eagerly opened their vast collection to my perusal. Over a two-year period, I visited their Research Department and culled through hundreds of photographs, carefully choosing those that I felt represented the Ford design department and Ford automobile design during E.T. Gregorie's tenure. This task was more difficult than it sounds. When Henry Ford ordered his company photographers to roam the company and take photographs, no rhyme or reason existed in their endeavors. On a single roll of film or during a photo shoot on a single day, they would take pictures of everything from flowers on the grounds of Greenfield Village to clay models in the design department. To make matters more difficult, the photographs were cataloged by sequence rather than by subject! Therefore, as I searched for the rare scene of the design department, I had to weed through folder after folder of unrelated images. I had to be careful as I searched, because one precious image of a clay model could be sandwiched among dozens of images of furniture, flowers, or farm equipment! Going through the thousands of images reposited at the Ford Archives is not for the impatient.

Eventually, I accumulated a couple hundred photographs. With copies in hand, I took them to Mr. Gregorie, and we discussed each picture and each design in detail while sitting in the sun room of his Florida home. I sat in a flowered rattan chair, and Mr. Gregorie sat on a couch. He was wearing shorts and a lightweight shirt, open down to the third button, and penny loafers with no socks. His red hair was now white and thin, and he had long white sideburns. He still sported a pencil mustache, just like he used to have while working for Ford. Occasionally, he would light a stubby cigar as he described how he designed a particular

Ford, Mercury, or Lincoln. Mr. Gregorie had never devoted so much of his time to recounting his years at Ford, but he seemed to enjoy telling me the stories as much as I enjoyed hearing them. Fortunately for me and for posterity, the 90-year-old designer recalled dates, names, and events surrounding his designs as if they had occurred yesterday. These discussions were never dry or technical; rather, they were always lively and highly informative. Mr. Gregorie is very articulate, and when he combines his mastery of the English language with his keen wit, the result is storytelling at its finest.

"I have to think back on some of the changes we made and the reasons why we made them and so on," Mr. Gregorie told me. "After all, seventy years is quite a long time! Boy, oh, boy, it makes you feel old! I have to kind of whack, whack the old memory! Back then, if I was told that I would be looking at cars seventy years from now and discussing a particular car and the lines on it, I would have looked at them funny."

Between stories, we stretched our legs by walking to the other side of the sun room, where a multitude of photographs hang on the wall. Some date back to the 1930s and are photographs of cars that Mr. Gregorie designed; others are photographs of the many boats he designed over the years. From the time he retired from the Ford Motor Company until only a few years ago, Mr. Gregorie and his vivacious wife, Evelyn, lived on yachts. In summer, they spent their time in and around Chesapeake Bay; in winter, they travelled up and down the Florida coast. "We wanted to stay out on the water longer," Mr. Gregorie lamented, "but it became too crowded and too rough." I could tell he missed the water very much.

When Edsel Ford appointed E.T. Gregorie the chief designer of the Ford Motor Company, Gregorie began staffing his department with young graduates of the Henry Ford Trade School and Ford Apprentice School. Mr. Gregorie was only twenty-seven years old at the time, but the men he recruited from these institutions were only eighteen or nineteen. Many remained with Ford their entire careers; others left the company when America entered World War II. Those who are still around told me marvelous stories about Mr. Gregorie, Henry and Edsel Ford, other Ford executives, especially Charley Sorensen and Pete Martin, and the often whimsical, occasionally dramatic, interactions of this group of men and the design of Ford products.

John Najjar, one of the original designers in the department, has kept tabs on many of his early co-workers. Through John, I became acquainted with Tucker Madawick, Ross Cousins, and Emmett O'Rear. Dave Crippen led me to Bud Adams and Benny Barbera, and through Dave's assistance I met Ralph Roberts, former head of the design staff of Briggs Manufacturing. Now more than one hundred years old, Roberts assisted Edsel in designing some of the Fords prior to Mr. Gregorie being appointed design chief of the Ford Motor Company.

In addition to the marvelous interviews I conducted and the photographs from the Ford Archives, I also studied the documented interviews of automobile designers in the Edsel B. Ford Design History Center. From this collection, I used the interviews of Holden "Bob" Koto, George Walker, John Najjar, Tucker Madawick, Franklin Q. Hershey, Rhys Miller, G. Eugene "Bud" Adams, and, of course, E.T. Gregorie.

Although Mr. Gregorie was the design chief at Ford for ten years, I could find only three photographs of him in the Ford Archives. Fortunately, he shared his personal collection with me. That collection included precious photographs of Mr. Gregorie when he was a young designer at Ford, his automobiles, the continental cars he designed for Edsel Ford, and the early Ford design department. After many letters and numerous phone calls, I eventually obtained from a private collection the one extant photograph showing the great designer in the design department.

As I worked on the 1946 Ford that my father had bought for me for $59.00, who would have guessed that I would eventually talk to the creator of this and so many other magnificent Ford automobiles? I hope that other Fordophiles and automobile enthusiasts will enjoy reading the stories and admiring the photographs in this book as much as I enjoyed compiling them.

Henry Dominguez
San Francisco, California

CHAPTER ONE

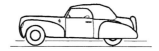

MR. GREGORIE AND MR. FORD
Designer and Patron

When E.T. Gregorie left the Ford Motor Company in 1946, America was entering a new era of prosperity. America and its Allies had just won World War II, and over the next twenty years, America's appetite for automobiles would gobble up more than 120 million vehicles, almost three times the number sold in the previous two decades. Ford Motor Company would be a major contender, selling an average of almost two million cars and trucks per year between 1946 and 1966.

E.T. Gregorie never knew such dizzying volumes. He joined Ford Motor Company during the depths of the Great Depression, when the company had slipped from first to third place against General Motors and Chrysler, both of which were making great strides in engineering and styling. Under the direction of Henry Ford, the company had been slipping precipitously behind its rivals in these critical areas since the mid-1920s and would have to make monumental steps in both areas to remain competitive. Henry Ford was never one to shy away from developing new mechanical features for his automobiles; rather, he was merely dilatory in introducing them. His son Edsel had the makings of one of the most talented design executives in the industry. Edsel had been toying with automobile styling since the early 1920s and would become the company's prime mover in implementing fresh body designs.

To help thwart the rising competition in engineering, Henry began working on his last mechanical triumph, the low-cost V-8 engine. Simultaneously, Edsel began working on a new body design for his father's mechanical innovation. By the end of 1931, the development work had been completed on what would become one of the most handsome and popular models offered by the Ford Motor Company. It was also the year in which the company hired its first true automobile designer.

Eugene Turenne "Bob" Gregorie became the styling genius at Ford Motor Company. During his years with Ford, he was responsible for the design of more than eight million Fords, Mercurys, Zephyrs, and Lincolns between 1931, when he started with the company, and 1946, when he retired to Florida. Although General Motors, under the design aegis of Harley Earl, produced substantially more cars than Ford during this same period, Gregorie's designs have withstood more than six decades of scrutiny. Indeed, the man Harley Earl so hastily let go in 1929 became his most artistic rival only seven years later.

Gregorie is a New Yorker, first and foremost. Although he spent most of his life in the Midwest and Southeast rather than on Long Island where he was born, he has retained a hint of the New Yorker accent and brashness. Yet, he can be as personable, warm, and witty as a Southern gentleman. Gregorie is well educated, extremely articulate, and highly confident. While at Ford, he could as easily talk shop with the "boys" as offer a compelling argument to senior executives at Ford concerning why one design was better than another. Although his design talent and eloquence earned Gregorie the top design job at Ford, his audacity and confidence enabled him to keep it for as long as he did.

Tall and lithe, with vivacious blue eyes, carrot-colored hair, and a mustache long enough to put little twists at each end, Gregorie himself was as stylish as his designs. In the dark-suited empire that was Ford in the 1930s and 1940s, he stood out in his sporty English tweed and seersucker suits and his buckskin suede shoes. On his $300-per-week salary, Gregorie could not afford the expensive, handmade suits worn by Henry and Edsel Ford; however, he did a commendable job in achieving that "English-cut" look from ready-to-wear suits that he bought for a meager $24.00. His master illustrator, Ross Cousins, remembers that Gregorie "would go to one of those low-priced suit places and pick something off the rack and have it altered a little bit, and you'd swear he got it at Brooks Brothers."

Gregorie often wore yellow, blue, or pink Oxford shirts with exactly one inch of sleeve extending below the suit coat, his pants hanging long enough to expose the color of his socks, which always matched the color of his shirt. "His pants always looked too short!" recalls Placid "Benny" Barbera, one of his designers. "He always dressed nice, but that was the only thing the guys used to look at him and say, 'Look at how high his pants are!' They were extra high, always an inch or so above his shoe tops. Sometimes they seemed shorter than that." One day, Gregorie would sport a narrow knitted tie and on another day a bow tie, but he always had a handkerchief jauntily positioned in the lapel pocket of his suit coat. In winter, he wore a raccoon coat. "In this life," Gregorie explained to one of his designers, "you have to find a way to attract attention, and *then* have something to sell." What Gregorie had to sell was confidence, audacity, and creative ability—attributes that Edsel Ford was desperately seeking in a man to head the design activities at Ford.

Gregorie's first love was boats, and he began his career designing them—not racy Chris-Crafts or streamlined sailboats, but husky, motor-powered yachts and seagoing vessels. He has owned a number of good-sized motor-powered yachts over the years, purchasing his first one in 1928. However, he would have been happier piloting one of the Ford Motor Company 400-foot Great Lakes freighters up the Detroit River than one of the sleek wind-powered yachts owned by Edsel Ford and other Detroit auto executives. This boating background gave Gregorie an eye for fine lines and clean, graceful shapes.

"Designing boats gives you a great sense of proportion and a sense of a beautiful line," Gregorie explains. "I can look at a line and tell whether it is fair, or if it should be tucked in here or tucked in there."

When Gregorie designed automobiles, his penchant for clean, flowing lines forced him to concentrate on the form, rather than on extraneous chrome strips and gaudy grille work as other designers did. "I've never cared for adornment on a car to give it style," Gregorie says.

Nor did his boss, Edsel Ford. Edsel also liked boats—not the kind Gregorie loved, but long and sleek race boats and graceful wind-powered yachts. In Edsel's mind, the sharp angles and uncomplicated lines of the yachts and speed boats he owned and admired were the epitome of gracefulness and simplicity, and he persuaded Gregorie to carry these nautical attributes into the cars he designed at Ford. For Edsel, as for Gregorie, the form of an automobile was paramount. Bumpers, grille work, and moldings had to be simple, clean, and delicate in appearance, and could not distract from the overall look of the car. Because of this affinity for boats, Edsel liked pointy shapes, and that is why most Ford products designed by Gregorie had sharp-looking grilles and pointed hoods, resembling upside-down boat hulls.

When Gregorie started working at Ford, the company did not have a design department and would lack one for another four years. It had a department called Body Engineering, which was responsible for body layout, tooling, and construction. By default, body engineering assumed the responsibility of body design as well, which explains why those old Model T's appeared so functional but sadly acetic. The 1928 Model A was the first Ford designed with a stylist's point of view, and it showed how stylists and engineers could work together harmoniously to effect a handsome but functional design. Many engineers were involved with the mechanical aspects of the Model A, but only three men were instrumental in its body design: Edsel Ford, Joe Galamb, and Amos Northup.[a]

Galamb, the chief body engineer at Ford, had been with the company almost since its beginning and was influential in designing the popular Model N and Model T. Northup, who worked for Murray Corporation of America, a major body supplier for Ford, was one of the earliest automobile stylists in the industry. Before the Model A, he had designed the Wills Sainte Claire; after the Model A, he designed the Hupmobile Century Eight, the Reo Royale, and the Graham Blue Streak. A few years later, Edsel, working with designers at Briggs Manufacturing, developed the refined 1932 Ford line. From that point, everything that Ford produced was designed by stylists, most notably Bob Gregorie.

The relationship that developed between Edsel and Gregorie was unique in automotive history. It was more than a mentor/student relationship or even that of a master and an apprentice. Their relationship was akin to that between a patron and an artist. Gregorie was the budding artist—young, talented, and full of promise and ideas. Edsel, the rich patron, derived pleasure simply by watching his young designer create. Gregorie leaned heavily on Edsel for support and protection, and Edsel depended on Gregorie for his creative abilities. The two men had such respect for one another that, in all the years Gregorie and Edsel worked together, Edsel always referred to Gregorie as "Mr. Gregorie," and Gregorie always referred to Edsel as "Mr. Ford."

Except for the 1932, 1935, 1936, and 1937 Fords (which were designed at Briggs Manufacturing under Edsel's watchful eye) and the custom-bodied Lincolns (which were designed and built by a myriad of custom coachbuilders), Gregorie designed or directed the design of every Ford, Mercury, Zephyr, and Lincoln produced by Ford Motor Company during the company's classic era of the 1930s and 1940s. Gregorie's sleek, unadorned designs were the envy of the industry, causing the boisterous Harley Earl, head of the formidable Art & Colour Section of General Motors, to lament on numerous occasions about the beautiful designs coming from Ford. "I never saw Harley Earl after I left General Motors," says Gregorie, "but he heard of me a lot. I was the only one who apparently gave him a hard time."

"Heard" is the correct word, for Gregorie accomplished all his work outside the limelight and in the seclusion of his Dearborn studio. His fellow designers, such as Harley Earl, John Tjaarda at Briggs Manufacturing, and Raymond Loewy at Studebaker, were flamboyant characters both inside their companies and in public. Compared to them, Gregorie was a virtual recluse. He and his

---

[a] Years later, Gregorie recalls seeing Northup give a speech at a meeting of the Society of Automotive Engineers (SAE). "He was up on the stage describing one of his automobile designs and made an embarrassing *faux pas*," says Gregorie. "He said, 'Now notice the mess of the car.' He meant to say, 'Now notice the mass of the car.' Everybody laughed, and I have never forgotten it!"

first wife, Gertrude, lived modestly by themselves in converted two-story stables on Grosse Ile, a little island on the Detroit River approximately fifteen miles south of the Ford design studio, far from haughty Grosse Pointe and Indian Village where many automobile executives lived. Gregorie's house was situated in the middle of the island on five acres of land. The upper floor was living quarters, and Gregorie converted the main floor, which previously housed horses, into a four-car garage. When Tucker Madawick, one of Gregorie's designers, visited him one day in 1940, he noticed Gregorie's Lincoln Continental prototype, a 1939 Ford convertible, and a 1939 Ford station wagon parked inside the garage.

For pleasure, Gregorie sailed his yacht, docked at the Grosse Ile Yacht Club, and rode his miniature railroad that he built around the circumference of his property. It was dubbed the Grosse Ile & Swamp Hollow Railroad and Gregorie installed a two-cylinder gasoline engine in a makeshift locomotive to pull several hand-built passenger cars. "We'd have cocktail parties and pull two or three carloads of people around and around the track," recalls Gregorie. "It was great fun."

John Hay, who worked in the model shop of the design department, remembers riding on Gregorie's miniature train one time. "I stopped by to see him when I went out to enlist in the Navy during the war," Hay says, "and he was out there on this train! So I hopped on, and we rode it around and around that doggone track!"

Harley Earl, John Tjaarda, and the other automobile designers knew Gregorie, but they saw him only rarely. Occasionally, they would see him at an SAE meeting, and usually they met him at the annual New York Automobile Show. However, in each instance, they would talk for only a few minutes. When Gregorie went to the New York Automobile Show, he was more interested in visiting his family on Long Island than hobnobbing with the other designers.

"There was a funny thing about me, the Ford Motor Company, and the relationship I had with Edsel Ford," explains Gregorie. "We operated in a separate little cocoon out in Dearborn. I never belonged to the Detroit Athletic Club, or the yacht clubs, or the other social clubs the other designers belonged to—they had a real clique! I stayed strictly out of that, because it's my nature. I love privacy. And Edsel Ford appreciated that. He was never a publicity hound, and he never encouraged that type of thing. As soon as he died, the company offered me memberships in four or five different country clubs. But that wasn't my nature to be that way, and that's the reason, I think, Edsel liked me. He was kind of a loner himself. We always enjoyed the idea that we worked together on the product for the company, and we didn't care for any outside interference. He immediately recognized my particular instincts and desires. That's the reason we got along so well. We appreciated each other's talents, and the product was pretty much worked out between the two of us. We did the work that the other companies needed thousands of people to do!"

Isolating the design department from the outside world was more than a desire for privacy. It was Henry's and Edsel's unwritten law that none of their top executives should fraternize with others outside the company. They didn't want their men being tainted by others' ideas, especially in the areas of engineering and design. Edsel wanted all design ideas from his design department to be strictly Ford-inspired. Gregorie wanted it that way, too, and he enforced that law on the men who worked for him.

"I didn't encourage any contact of our people with other manufacturers," Gregorie explains. "I told my staff, 'I don't want to see a line on any of these automobiles that smacks of anything anyone else has ever done.' I just had a pride that I was never going to put a line on a car that was obviously taken from another car. I thought that if I couldn't do something original, something that smacks of Ford design department, then I'd better get the hell out of it. That's the reason our cars had individuality."

Throughout Gregorie's tenure at Ford, engineering and production were paramount because Henry Ford wanted it that way. From the time Henry Ford created the company in 1903 until his death four decades later, engineering and production reigned. "The old man had absolutely no interest in the design and appearance of the car," explains Gregorie.

Henry Ford surrounded himself with rough and tough engineers and production men who ruled the mammoth Ford industries with fists of iron. Charles Sorensen, a tall and handsome Dane, was the most notorious of these men. Pity the man who looked up from the production line when Sorensen walked by, for Sorensen would fire him on the spot. Sorensen had been Henry Ford's right-hand man since the Model T days, and Henry had grown to appreciate him more than his own son.

Another one of Henry's lieutenants was the diminutive Pete Martin. "He was a rough, old son-of-a-bitch," says Gregorie. "If some guy looked at him cross-eyed, he'd yell, 'Go get your pay!'"

Joe Galamb was a Hungarian from the old country. Henry Ford liked Galamb because he was, in Gregorie's words, "a clever put-togetherer." Even after being in the United States for years, Galamb never entirely lost his thick accent. For example, when Galamb said "sheet metal," it sounded as if he were saying "shit-a-metal." Mimicking this, the men began calling him "shit-a-metal Joe"—behind his back, of course!

"Galamb was a nice-looking guy," explains Gregorie. "He had a gray mustache, sported fashionable clothes, and wore plenty of perfume. But he was an arrogant little man. Very pompous. He looked like a foreign diplomat and treated Henry Ford as though he were a king. He'd always bow to Mr. Ford when he saw him."

Galamb worked with Edsel on body design when Edsel assumed design responsibilities. However, when Gregorie entered the picture, Galamb had to relegate his responsibilities to the new chief designer, and he always resented Gregorie for that. Galamb may have resented Gregorie for taking over his design responsibilities, but if Galamb hoped to retain his job, he did not take steps in assuring his future. "The use of clay models and styling bridges were way over Joe's head," recalls Gregorie. "He'd only been familiar with paper templates and had never learned anything else. We were a few stages ahead of him."

The only educated man in the bunch was Larry Sheldrick, chief chassis engineer at Ford; however, he was no less churlish than the others. "He [Sheldrick] was very assertive, hard boiled," explains Gregorie. Sheldrick had kicked and scratched his way up to his position during the Model A days, and he would not readily relinquish any authority to a youthful upstart such as Gregorie. He had to work with Gregorie in developing new platforms; however, because Sheldrick was part of Henry Ford's "Russian Council," he resisted Gregorie's ideas whenever possible. "Oh, boy, Larry [Sheldrick] disliked me!" recalls Gregorie. "He made it tough!"

Fortunately, Gregorie was equally tough and never backed down from Sheldrick or any of the other "Henry Ford-men," most of whom were old enough to be his father. He had to hold his ground, for if he showed the slightest sign of weakness, Sorensen, Galamb, and the others would have run him out of the company in the same way that they had done with so many others before him. "I had to mix it occasionally with these people to get what we wanted," explains Gregorie. When the Henry Ford-men put up resistance about the curvature of a fender or the design of a grille, Gregorie would retort, "Design is a hell of a lot more important to customers than whether the car has a straddle-mounted axle pinion, or roller bearings instead of ball bearings in the wheels. They buy the car on appearance!" It took courage for a young man to stand up to Henry Ford's cronies in that way. "I just took the chance," Gregorie explains. "I thought, well, if I'm gonna move ahead, I'm gonna move ahead. But I'm not going to get stuck in an ice jam here."

Gregorie's audacity irked them all the more, but he knew they couldn't do anything about it as long as his patron, Edsel, protected him. In fact, Edsel enjoyed it when Gregorie got into these verbal fist-a-cuffs with his father's cronies. "During these heated discussions, Edsel let me do the dirty work," explains Gregorie. "He would stand back and give me a certain look, which gave me the reassurance that I was on safe ground—that he would come to my defense if needed."

Eventually, a truce of sorts was reached between Gregorie and Henry Ford's men, which lasted until Edsel's death. "Once those henchmen knew that I worked for Edsel Ford and was able to please him and get things done," Gregorie says, "why, we ran over them like a steamroller!" They realized that Gregorie was there to stay and that he was amenable to their suggestions when they ran into a problem building one of his designs. "They would come up to me and want to change a stamping or something to accommodate the interchangeability of a part or a joint line, or some damn thing," Gregorie says, "and we would work things out."

Gregorie gladly worked with the manufacturing men because he understood their plight. If they couldn't manufacture a certain aspect of one of his designs, then he did not pursue it. "In designing cars," he says, "you stay as close to a pretty concept as you can, but you have to compromise sooner or later and do a few things that might be forced by other factors." Gregorie was very pragmatic in this regard. "Hell, I'm no artist!" he exclaims. "I'm a ship designer. And I ran the design department like I would a shipyard or a boatyard—it was all business. When I hired men for the department, I told them, 'I don't want any affections of artwork. This is a business, and you're going to have to get your hands dirty.'"

Gregorie's greatest challenge was not Henry Ford's men or Edsel's propensity for sharp, pointy shapes. Rather, it was the Ford chassis, the foundation on which he had to develop his designs. From Gregorie's stylistic point of view, the Ford chassis was too high off the ground, its engine was placed too far forward, and the front axle was too far rearward to effect a well-proportioned design.

"It's the worst set of combinations a designer could be handicapped with," Gregorie exclaims. "It was intolerable to think how it disfigured the car, but the chassis layout was never taken into consideration as to the potential ill effect it had on the overall appearance of the car. You just simply had to make the best of it. If you attempted to pull the hood out, it would have been even worse."

Over the years, Gregorie tried desperately to convince Henry Ford—through chassis engineering—to simply move the front axle forward, if only a few inches, so that he could give the Ford a more balanced appearance. He actually demonstrated that the front axle could be moved without altering the frame rails by utilizing some ingenious brackets he designed. However, old Henry Ford would never concede, and Edsel always lost the argument whenever he broached the subject with his father. Henry already thought his cars had too much styling as it was, and he would not alter his precious chassis for the benefit of the design department. In fact, after Edsel's death, Henry Ford told Joe Galamb, "Joe, we have to go back to the old days, the Model T days. We've got to build only one car. There won't be any Mercury, no Lincoln, and no other car—just the one Ford. You can put any kind of body on there as long as it's the same chassis. Don't let anybody change that idea, either."

Fortunately, that didn't happen. Henry was merely reminiscing about the old days when he controlled every aspect of the company. However, those days were gone. By the time of Edsel's death, Henry was old and feeble, and company policy was being watched closely by the board of directors. They would never have agreed to such a scheme.

Moving the front axle forward would have balanced the front end of the car with the rear end and given the car a much better ride. "It would have improved everything," laments Gregorie. Instead, he was stuck with an antiquated chassis that was skewed forward, causing him untold grief in developing well-proportioned designs. "I don't care how good of a designer you were," recalls Gregorie, "you couldn't do much with the front end and the front fenders. I couldn't even do much with it."

Gregorie is much too modest. The chassis constraints notwithstanding, the front-end designs he created are among the most renowned in automotive history. We can only image what beautiful forms Gregorie would have developed if Henry Ford had just moved the front axle!

Another Henry Ford requirement that stymied Gregorie's artistic license was old Henry's insistence on ensuring that the basic Ford met the needs of American farmers. "We always had to keep in mind that Henry Ford considered the Ford a utility car and that farmers would want to go out in the field and pick one or two bushels of potatoes, or pick up a couple of milk cans from the co-op." Gregorie laughingly recalls. "That old Hungarian, Joe Galamb, would come in and say in his thick accent, 'Mr. Ford vood like to put some booshell baskets, booshell baskets in de trunk.'"

Bud Adams, who started as one of Gregorie's apprentices in 1938, recalled seeing the outline of a milk can, drawn in phantom, standing in the trunk on the body draft of the 1937 Ford. Such constraints forced Gregorie to design the rear end of Ford cars—especially sedans—with a high and expansive trunk area, a practice he was compelled to follow through the 1948 model year. However, unlike the chassis constraints, Gregorie was able to take advantage of these requirements, giving Fords that graceful fastback design for which they are now famous.

No other designer in the automobile industry during this time was subjected to such arcane thinking. Harley Earl, with complete support of his boss, Alfred Sloan, had successfully separated engineering from styling and was moving quickly toward lower, longer, and wider designs. Chrysler's Fred Zeder, Owen Skelton, and Carl Breer, although engineers, realized the importance of styling and worked feverishly on streamlined designs. For the most part, Gregorie was stuck with antiquated management thinking and a chassis that was not very different from that of a 1908 Model T. This dichotomy underscores Gregorie's talents all the more. Only he could have worked in such a design-adverse environment and still create beautiful, graceful automobile designs.

If Gregorie had done nothing but protect Edsel's fledgling design department from Henry's henchmen, he would have accomplished what Edsel needed. However, Gregorie did much more than that. He staved off the affronts from Henry Ford's men, and he created a design department within Ford Motor Company that rivaled any in the industry. That accomplishment took confidence, patience, and temerity.

"Gregorie was aggressive and ambitious," explains Ross Cousins, "and he was determined to make the design department successful." John Najjar, another early designer hired by Gregorie, concurs. "It was quite an accomplishment," Najjar says, "for a young man starting an in-house design department from scratch, having to gather all new personnel, carve out a work space, in what must have been a hostile environment."

The hostile environment and antiquated thinking aside, "Gregorie had it knocked!" exclaims Placid "Benny" Barbera. "Mr. Gregorie didn't come into the design department in the morning too much, especially in summer—he was on his boat! He spent many summer mornings cruising up and down the Detroit River, presumably clearing his head and thinking up new ideas. He made it to work just before Mr. Edsel Ford came in after lunch. Oh, Gregorie was a slick operator, a sharp guy, a good, fast talker, and a fast sketcher. Edsel Ford gave Gregorie a free reign, and he took advantage of the free reign. Gregorie was no dummy."

Even when Edsel was around, Gregorie would walk over to the company garage, sign out a shiny new Ford roadster, and take long drives in the country. That was unheard of at Ford Motor Company prior to Gregorie joining the company. Any other man in the Ford organization who attempted to do such a thing would have been fired on the spot. But not Gregorie.

"He tried to remain somewhat aloof," explains Ross Cousins. "He would not punch the clock. He was determined not to follow those kinds of rules. He wasn't about to, and Edsel didn't want him to."

Edsel understood that the creative mind must be free from the often stifling and mundane office environment, and he allowed his chief designer these indiscretions. Edsel's support and generosity did not go unrewarded. From his new design department came some of the most innovative and beautiful designs in automotive history.

"I can tell if a line is good," explains Gregorie, "and as I walked around the clay models, I would tell the 'boys' to add some clay here and take some clay off there, until every surface line looked right to me. Looking at all those designs today that we did back then, I still wouldn't change a line on any of them."

Gregorie was the first to use styling bridges in developing clay models, a technique heretofore used only in building ships. Previously, the body engineers would painstakingly transform a three-dimensional wooden or clay model into a two-dimensional engineering drawing. With Gregorie's technique, he would merely hand them a book containing page after page of coordinates taken from the modeling bridge! Another unique design tool that Gregorie used was a twelve-foot-high elevation tower from which he could analyze body shapes. "Viewing a car from above," he explains, "gives you a more accurate perspective of the design of the car. It allows you to see how all the different surface lines blend together."

While eating a candy bar one day after lunch, Gregorie noticed that the foil wrapper from the candy bar resembled real chrome when stuck to a clay model, and the industry has been using aluminum foil to simulate chrome plating on its mock-ups since then. Gregorie also initiated the first two-spoke steering wheel and was first to develop the horizontal grille, a design cue that set the automotive world on end.

Unlike many design directors at that time and since then, Gregorie did a great preponderance of the design work himself. "I was pretty handy with a pencil," he proudly explains, "and would do a lot of the pencil work myself. I mean, I could draw automobiles from every angle, and I always loved to draw perspectives. All the designers at Briggs, for example, could draw nice side views, but they couldn't draw perspectives. Gosh, I could draw cars up on grease racks, showing the underside and the whole damn thing." Gregorie is one of the rare few who can honestly say, "My designs are my designs. I did the designing myself."

Automotive historians have listed the 1936 Zephyr, 1938 Zephyr, and 1939 Continental as Gregorie's greatest achievements. Frank Lloyd Wright called Gregorie's Continental "the most beautiful car in the world." In 1951, reflecting on the fact that automobile design was an art form, the Museum of Modern Art called Gregorie's 1936 Zephyr "the most successfully streamlined car in America." However, Gregorie believes that the 1933 Ford and 1949 Mercury were his best work—the 1933 Ford because it was almost perfectly proportioned, and the 1949 Mercury because it was the first modern car Gregorie was able to design from a clean sheet of paper, one that did not have to accommodate either "booshell baskets" or milk cans in the trunk. However, Gregorie mentions these two cars only after being asked, more than fifty years later, which of his designs he likes best. During the years that his multitude of Ford, Mercury, and Lincoln designs were coming off the assembly line by the thousands, Gregorie did not give the subject much thought. To him, designing cars "was just an everyday happening," he now says nonchalantly. "They were just cars. There were so many steps along the way—designing the car, the clay model, the prototype, and so on—that by the time the car was coming off the assembly line, it was nothing new. The excitement of the design had been diluted over the full length of the project."

When Edsel Ford died in 1943, Gregorie lost his patron and protector, and he left the Ford Motor Company a short time later. He tried to remain longer because he loved his work, but working for Henry Ford II, Edsel's eldest son, was not the same as collaborating with Edsel. When young Henry hired Ernest Breech from General Motors in 1946 to help him run the company, it was too much for Gregorie to tolerate. Thus, Gregorie left Ford at the age of 38, never to return to automobile styling.

We can only speculate on how this believer of "the perfect form" would have mitigated the chrome colossuses produced in Detroit in the 1950s, had he remained in the industry for another decade. Harley Earl's tail fins and Virgil Exner's glitzy behemoths were the antithesis of Gregorie's penchant for clean, uncluttered designs. When asked what he might have done to thwart these atrocities, Gregorie replied, "Oh, I would've had something in my hip pocket."

This is the story of E.T. Gregorie and the graceful automobile designs that he pulled from his pocket for Edsel Ford.

CHAPTER TWO

FORD HIRES A DESIGNER

"I've Been in Body Design at Brewster"

E.T. Gregorie was born on October 12, 1908, in Long Island, New York, the first son of Eugene T. Gregorie and Alva Palmer Gregorie. (Figure 2-1) Being a "junior," he was called "Bob" by his family to distinguish him from his father, and the name has stuck since then. Many people incorrectly pronounce his last name as GREG-er-ee or Greh-GORE-ee, but he pronounces it GREG-ree, with the *o* silent. "Sometimes," Gregorie sneers, "I think about just getting rid of the *o* and spelling my name G-r-e-g-r-i-e. That would force people to pronounce it correctly."

Gregorie's ancestors immigrated from Scotland in the eighteenth century, settling in Virginia, where they were merchants, importing products from Britain and "dabbling in the slave trade." His uncle, James Gregorie, was a professor of astronomy at the University of Glasgow in Scotland.

Gregorie's father, a successful businessman and shrewd investor in the stock market, was said to have possessed the same rare talent as Henry Ford in being able to distinguish one steam engine from another simply by its sound. Mrs. Gregorie was very imaginative and an accomplished painter.

Gregorie possesses both the mechanical horse sense of his father and the artistic ability of his mother. In later years, the engineers at Ford were amazed that he could talk horsepower and torque as well as he could sketch a design. "That came from my experience as a naval architect," says Gregorie. "Naval architecture amounts to part engineering and part art, and I knew as much about the mechanical parts of the car as I did about the appearance of it."

Gregorie attended private schools in Virginia, and he spent much of his free time (and some of the time when he should have been in school) around the boat docks of Long Island, Maryland, and Virginia. This started Gregorie's love affair with the ocean and boats, which continues to this day.

Gregorie realized early that he had artistic skills, but he can't explain how they developed or why he possessed an intrinsic ability to discern a classic line from a mediocre line. "It's just one of those things," Gregorie says. "It just all fell together. It's like a guy who has an eye for a good-looking gal's leg, you know. You kind of form an opinion. I began drawing imaginary boats and cars when I was about fifteen. I started drawing boats because

Figure 2-1. E.T. Gregorie, shown here in the summer of 1940, usually dressed more nattily than this. However, he was warned that he would have an official company portrait done and thus wore a conservative business suit that day. (Photo from the Collections of Henry Ford Museum & Greenfield Village)

that's what I grew up around. And then we had cars, sporty-looking cars, and I began drawing them. But my drawings weren't based on any actual concepts. You might say they were based on the existing good-looking cars, good-looking boats—the ones that impressed me. Oh, I can recall cars that I liked the looks of, like the old Marmons, some of the Packards, Cadillacs. These were cars I considered favorably, and I tried to emulate them."

In those early days of industrial design, there were no schools in which young men could study design. They simply learned it along the way, and that was how Gregorie developed his skills. In 1926, he took his inherent artistic talent to Elco Works, a yacht-building company in New Jersey. There he began developing his design skills as a marine draftsman under the watchful eye of William Fleming, the chief naval architect in the company.

"I learned an awful lot from Bill Fleming," explains Gregorie. "He always had a pencil in hand, and while he described something to me, he'd make a sketch in perspective at the same time. From him, I learned how to become proficient at doing the same thing, which helped me later in the automobile business. I liked his work, and I liked working for him. He was tall and gangly and wore thick glasses, and he was very amusing. He'd come to work wearing a winged collar, as if he were going out to a cocktail party, and he'd leap over the small partitions we had in the office. He was full of life, and very talented."

Despite Gregorie's enjoyment in working for Bill Fleming, the long daily commute finally got the best of Gregorie. "It took me three and a half hours each way to get to work!" Gregorie explains. "I rode the Long Island commuter train to New York, and then made two changes on the subway down to the ferryboat over to New Jersey. Next, I took a 45-minute steam-train ride down to Bayonne. I put up with that for about a year, and then I couldn't take it any more—it was getting me down. Plus, it cost me $15 a week to go back and forth to work! During the summer of 1927, I went on a camping trip to Pennsylvania with a couple of friends, and that's when I decided to quit."

After Gregorie left Elco Works, he never saw Fleming again. "He would have been proud of me," says Gregorie. "He would have felt very good about what I wound up doing at Ford. I intended to stop by and see him in one of my sporty cars that I had designed at Ford, but I never got around to it."

With his experience from Elco Works, Gregorie quickly landed a job at Cox & Stevens, another prestigious naval architectural firm, located on Fifth Avenue in Manhattan. "It was much easier to get to," says Gregorie. "I only had to take one train from Long Island and get off at Penn Station and walk up to the office." Regarding his experience at Cox & Stevens, Gregorie says, "I was able to work with some very renowned marine designers, like Philip Rhodes and Daniel Cox. It was excellent training."

Gregorie merely refined his design skills at these firms, for he already had the innate ability. "You've got to have a good eye for an attractive boat, a nice beautiful sheer line, which is the profile of the hull," Gregorie says. "That's the first place that shows up from an amateur's attempt to design a boat. If you start with a sick-looking boat profile line, you might as well quit. It's a question of proportion: the size of the deck house, the rake of the stem, the rake of the stern, the curvature of the stern. (Figure 2-2) A good designer naturally arrives at a good compromise on those things." Gregorie always arrived at a good compromise.

To survive at these companies, the young Gregorie had to be not only a talented and resourceful designer, but also able to articulate his design ideas to customers who had more money than design sense. Gregorie already had a sense of gab, but during the years he worked at these naval architectural firms, he developed his verbal skills even more. By the time he left Cox & Stevens in the fall of 1928, he could paint a picture in the mind's eye as poignantly as Robert Frost, practically selling a customer on a design before drawing a single line. His eloquence became even more important in the automotive industry. John Najjar says that Gregorie could "plan, present, and follow through on an idea better than anyone else."

Automobiles intrigued Gregorie's father as soon as they began appearing on public streets. He eventually owned a number of them, from luxurious Delages to sporty Simplexes, during the years his son was growing up. One car that Gregorie fondly remembers was his father's 1912 chain-driven Mercedes.

"It was a big red touring car," Gregorie says, "with right-hand drive, wire wheels, big spares on either side, and big brass levers on the side of the driver's seat. It had no starter. To get it started in winter, you had to prime it with ether, release the compression, hand-crank it to get the flywheel going, then switch on the compression. It even had water-cooled brakes. It was an Alpine model—meant to be driven in the Alps—and it had a water tank on both sides, each about three feet long and eight inches in diameter. They held about eight gallons of water, which circulated around the brakes through tubing. So when the brakes got hot, a thermo-syphon action developed, which circulated the water in the tubing, keeping the brakes cool. We used to

Figure 2-2. This 75-foot marine photographic vessel was designed by Gregorie in 1966, but it is typical of the graceful yachts he designed in the late 1920s before turning his creative skills to automobiles. (Photo courtesy of E.T. Gregorie)

take long drives with that car on Sunday afternoons. We'd go sixty or seventy miles out, going forty-five miles an hour. I loved to hear those old chains singing—*ying, ying, ying, ying!*"

It wasn't until Gregorie worked at Cox & Stevens that he began to take greater interest in automobile design. He would have loved to have spent his life designing yachts but soon realized that the demand for yacht designers was limited by the number of wealthy patrons who could afford such crafts. This soon led him away from naval architecture to automobile design.

Late in 1928, Gregorie took several of his most interesting automobile body designs to Brewster & Company (then owned by Rolls-Royce of America), and he showed them to the general manager, John Inskip. After looking at Gregorie's work, Inskip hired Gregorie immediately.

Brewster had been a custom carriage maker since 1810 and began designing and building custom automobile bodies at the turn of the century. By the time Gregorie started working there, Brewster was one of the premier custom-body houses in the country and the most prominent training house for many budding designers. Ray Dietrich, perhaps the most famous of all the old-time custom-body designers, got his start there, as did Thomas Hibbard. Beginning by humbly making design sketches of custom bodies for all types of automobiles, Gregorie developed his unique skill of adapting graceful nautical lines to automobile design while at Brewster. Gregorie was not totally sure that automobile design would be his forte when he joined Brewster, but by the time he left a year later, he realized that it was the field he wanted to follow.

In late summer of 1929, business at Brewster suddenly dropped. As in the yacht business, Brewster's customers were financiers and other wealthy people, and for some unknown reason, they stopped placing orders for custom-made automobiles. It was as though the drop in business were a bellwether of the tough economic times that lie ahead. "I wasn't laid off," says Gregorie, "but I thought it was a good time to think about moving west, so to speak. I intended to do that anyway. The pay at Brewster's wasn't that great, and the lack of orders coming in kind of made up my mind. But it was good training. I learned a lot there." Upon leaving Brewster, Gregorie packed his bags and headed for the gigantic production companies in Detroit, a trip that would change his life forever.

In the last few weeks before the Great Depression, Detroit was a denizen of activity. The Big Three—Ford, General Motors, and Chrysler—along with a number of independent automobile manufacturers, where churning out automobiles at the combined rate of 300 per hour. By this time, dependability and the techniques of mass production had been all but conquered. However, in terms of design, only two companies had created departments to work on this increasingly important aspect of automobile manufacturing: (1) General Motors, with Harley Earl's Art & Colour Section, established in 1927, and (2) the design group at Briggs Manufacturing headed by Ralph Roberts, also established in 1927. At Ford Motor Company, Edsel Ford had the goal in the back of his mind of establishing a design department, and it was only a matter of time before he would implement it. His first step was finding a man capable enough to head a design department within a company that was first and foremost production oriented, and whose top executives, headed by Henry Ford himself, would do everything in their power to make the department fail. At twenty years of age, Gregorie could hardly have anticipated such a hostile environment, but having the fortitude to strike out on his own to work in the biggest industry in the country, in the youngest and most precarious aspect of the business, says a lot about the man and his ambitions. To survive in this setting, he would have to call on every mechanical aptitude and artistic endowment that he inherited from his parents.

Detroit was Gregorie's destination, but he made stops at two automobile manufacturers along the way, "just to hedge my bet." He first stopped at the H.H. Franklin Manufacturing Company in Syracuse, New York, and talked with its chief engineer, Kenneth Haven. "They were in a bad way," explains Gregorie, "and about to throw in the towel," so they could not afford to hire a new man.

However, Haven told Gregorie that Ray Dietrich had been there only a few hours earlier and showed him some designs Dietrich had done for them. "When you get to Detroit, why, you look up Ray," suggested Haven, "and maybe you can work out something with him."

Gregorie did not have to wait until he reached Detroit to see Dietrich; he met Dietrich at the train depot while buying a ticket to Buffalo on the New York Central. "My family had owned Pierce-Arrows over the years," says Gregorie, "and I thought I'd go over there and tell them my story." While in line to buy his ticket to Buffalo, Gregorie overheard a tall man in front of him, who was holding a portmanteau of art, ask the ticket agent, "Do you have a reservation for Raymond Dietrich?"

After the train was underway, Gregorie found Dietrich in the parlor car and introduced himself. "I'm on my way to Detroit," Gregorie explained to the dashing designer. "I've been in body design at Brewster." Being an old Brewster alumnus himself, Dietrich immediately warmed to Gregorie.

"Sit down, sit down," said Dietrich, as he pulled a flask of gin from his hip pocket. "Would you care for some?" Gregorie politely declined, which he later regretted. "It was Prohibition, you know," he says. "But I was too young to consider the value of it. Anyway, we sat together in the parlor car as far as Buffalo—Ray was going on to Detroit—and we had a very nice talk about automobile design."

It was clear during their discussion that Dietrich would not be able to help Gregorie land a design job, but as the train pulled into the station at Buffalo, Dietrich asked Gregorie if he knew where he was going to stay when he reached Detroit. Because Gregorie did not know, Dietrich mentioned the Lewis Hotel located on Woodward Avenue near the General Motors Building. "They usually have rooms available there at a fair price," he said. "I have a room there, and I know a number of GM designers who stay there also." Gregorie thanked him, and the two designers parted.

Pierce-Arrow was one of the most prominent automobile manufacturers in the high-class field and equipped many of its fine chassis with custom-body work from all the major custom-body houses in the country. We can only speculate on the beautiful bodies Gregorie would have designed for Pierce-Arrow had they hired him. "They were about to fold up, too," Gregorie explains. "So I bought a five-dollar ticket to Detroit on *The Greater Detroit*—a big old wooden Detroit & Cleveland side-wheeler steamer. Those big side-wheelers were about three hundred feet long, and they were fast. They'd run eighteen, twenty knots, and they'd creek and groan with all the old woodwork in them. I think it was about two hundred and sixty miles up Lake Erie to the mouth of the Detroit River. They'd leave about six o'clock in the evening, and by six o'clock the next morning, they were going up the Detroit River. Anyway, I took the steamer that night and arrived in Detroit the next morning, Saturday, at seven o'clock. I remember it well, because it was my twenty-first birthday." As Ray Dietrich had suggested, Gregorie grabbed the northbound Woodward streetcar, got off at West Grand Boulevard, went to the Lewis Hotel, and got a room.

On Monday morning, Gregorie went directly to the General Motors Building and showed his portfolio of designs to Howard O'Leary, Harley Earl's front office man. In the portfolio were a number of automobiles that Gregorie had created. "They weren't of any particular automobile," says Gregorie. "It was just a packet of sketches to show that I knew what an automobile was and could draw them. It would be typical of what any designer would carry when he was out peddling his wares."

"In those days," Gregorie explains modestly, "O'Leary would hire anybody who could draw an automobile, and see if they would work out—kind of a screening process." Nevertheless, O'Leary was genuinely impressed with Gregorie's work and gave him a job in Earl's Art & Colour Section.

Never shy, Gregorie quickly made acquaintances with several young designers in the department. One of them was Franklin Quick Hershey, who had recently come from Los Angeles, where

Figure 2-3. Here Gregorie is shown behind the wheel of his 1925 Citroën in Detroit. Owning such an exotic little car in the heart of Ford-land was enough to make this car stand out in the crowd. However, Gregorie thought the original fenders on the Citroën were too skinny, and thus he adapted a set of Model A Ford fenders—the front ones tied together with a turned aluminum apron—to the little car. "That made all the difference in the world," says the designer. "Those fenders, which cost me only twenty dollars for the whole set, gave the little car more presence." (Photo courtesy of E.T. Gregorie)

he had been designing custom bodies for the Walter M. Murphy Company. Hershey later headed the design studios at Pontiac and then at Cadillac, and eventually he took over the Ford design studio when Gregorie's department was split into separate divisional studios many years later. However, at the time he and Gregorie met, Hershey was moving from an apartment in a large mansion not far from the Lewis Hotel, and he offered it to Gregorie, who took it. "A nice old gal and her daughter ran the place," Gregorie says, "and on weekends, some of the fellows from work and I would hold all-night poker games upstairs."

A few weeks before Gregorie moved into his new apartment, the stock-market crash of October 29 hit Wall Street. Now it was early December, and Gregorie decided to return to Long Island by train to get his belongings and his car, a 1925 Citroën cabriolet. (Figures 2-3 through 2-5) This little vehicle was Citroën's "bread and butter" car, just as the Model T was Ford's. Similarly to the Model T, it had no starter, no water pump, no heater, and only one door on the right side. "It was a real adventure driving that thing to Detroit in winter," recalls Gregorie. "When I left, it was very cold, mean. I had to take newspaper and caulk the cracks between the windows and canvas top to keep as much cold air from coming in as I could." Occasionally, Gregorie would stop at roadside stores to buy hot coffee or tea and warm himself at their wood-burning stoves.

A few miles beyond Hawley, Pennsylvania, Gregorie's car broke a valve spring. If it had not been for the inventiveness of a local mechanic—who adapted a valve spring from a wrecked 1924 Buick to work in the engine of the Citroën—Gregorie surely would have had to abandon his little French sports car in the snow. "He put the valve spring in, and the whole thing was a half hour and two bucks, and I was on my way again."

From that point, Gregorie made good time until he passed Wayland, New York, a small town sixty miles south of Rochester. It had been snowing all day, and the snow was accumulating on the ground but melting on the road. Around five o'clock in the

Figure 2-4. A dapper Gregorie stands beside his 1929 La Salle coupe in the winter of 1932. This photo was taken at the yacht club where Gregorie kept his boat. (Photo courtesy of E.T. Gregorie)

evening, the wet roads had turned to ice, and Gregorie lost control of his Citroën going around a curve. After spinning around three or four times in the middle of the road, the car careened into a cornfield and landed on its left side. Gasoline began pouring from the gasoline tank located in the cowl. Uninjured, Gregorie quickly shut off the gasoline valve located under the dash and climbed out the right-hand door. Two farmers, who had been behind Gregorie in a Model A Ford, stopped to help him. The three men uprighted the little car and, using a large rope the farmers had in the back of their Ford, towed the little Citroën to a garage at Wayland. Nothing could be done to repair the car until the next day.

Figure 2-5. Gregorie's mother, Alva, sits behind the wheel of a 1930 Whippet roadster. Gregorie bought this car for himself but was unhappy with the curvature of the front fenders. Using a pair of metal snips, he trimmed off the front portion of the front fender. He also added a machine-turned, polished aluminum splash apron, and he installed aluminum discs over the wire wheels to give the car a more sporty appearance. "My mother liked the looks of the car so much," recalls Gregorie, "that she bought it from me." (Photo courtesy of E.T. Gregorie)

The farmers told Gregorie that at the edge of town an old woman had a house with an extra room, and he could probably stay with her. "So I stopped by there and rented it," says Gregorie.

Fortunately, the Citroën had only a bent rear wheel, which the garage readily straightened the next day. However, it had snowed heavily all night, and by the next morning the snow was four feet deep on the roads. Gregorie could do nothing but remain in the old woman's house until the weather broke. "It was one of those experiences I look back on fondly," says Gregorie. "She fixed me nice meals, and by the time I left there, three days later, I owed her about six dollars."

When he was finally able to get back on the road, Gregorie made good time until he reached the eastern edge of Lake Erie. "There was a strong westerly wind blowing off the lake, which held back my little Citroën," Gregorie recalls. "I had to keep it in second gear to keep from going backwards!"

When Gregorie had left Long Island, he anticipated taking only two or three days to drive to Detroit. With the breakdown in Pennsylvania and the accident in New York, combined with the severe weather, Gregorie arrived at his apartment building ten days later. However, that was less surprising than what he found when he went to work the next morning. The economy had seriously worsened since the stock-market crash, and, uncertain of the future, General Motors began laying off its recent hires. Gregorie was one of them.

E.T. Gregorie and Harley Earl would have hit it off if given the chance. Gregorie would have held his own against the tall, handsome, and boisterous Earl, who would have respected him for that. Also, the designs that Gregorie surely would have developed for Earl would have endeared him to the first design chief of General Motors. However, as it happened, Gregorie was on the street, looking for a job rather than creating designs for Harley Earl.

Undaunted and confident in his design skills, Gregorie immediately sought a design job at the other manufacturers in Detroit, including the Lincoln Motor Company. Unfortunately, Lincoln was reeling from the Crash in the same way as all other automobile companies and was not in a position to hire any designers. Thus, Gregorie fell back on his nautical design skills and landed a job at the Detroit office of his old employer, Cox & Stevens. There, he helped finish the design of a 126-foot yacht for multimillionaire William Rands, founder of Motor Products Company, a major automotive parts supplier. That job lasted until midsummer of 1930. When it ended, Gregorie had to return to Long Island but did not abandon hope of working in the Detroit automobile industry. "I knew I had a future there," Gregorie says.

Gregorie spent the remainder of the summer skippering a large yacht for a wealthy friend. Then he went with his family to spend the winter at its South Carolina cottage. Shortly before New Year's Day, 1931, Gregorie received a telegram from Lincoln's chief body engineer and Brewster alumnus, Henry Crecelius, whom Gregorie had met while soliciting work at Lincoln. The telegram said that Lincoln needed a body designer (a directive given by Edsel Ford himself) and that the position could be his if he wanted it. Gregorie could start immediately. "Report at Ford Motor Company Engineering Laboratory, Dearborn," the telegram concluded.

Without hesitation, Gregorie said good-bye to his family once again and headed for Detroit. His uncle drove him to the railroad crossing in Yemassee, where the Atlantic Coast and the Charleston & Western Carolina railroads crossed each other. "It was a cold night," recalls Gregorie, "and I waited up in the signal tower, where the signalman kept a little coal stove going to keep warm." When a northbound Atlantic Coast fruit train came by at two o'clock in the morning, Gregorie caught it and got off at Charleston. From there, he boarded a Clyde Liner steamer at one o'clock in the afternoon and headed for New York City.

Tired of pacing the deck of the ship, Gregorie decided to pass some time by allowing a gypsy on board to predict his future. "I can see a small, dark-haired man," she said as she read her tea leaves. "And I can see a long, low, white building. He's going to be an important part of your life." Gregorie did not give much thought to what the gypsy was telling him—he was merely passing time anyway—but her predictions were uncannily prophetic.

After disembarking at New York, Gregorie headed to Grand Central Station and caught the *Detroiter* to Michigan. When he arrived in Detroit, he hailed a taxi and asked the driver to take him to the Ford engineering laboratory in Dearborn. As the taxi turned left onto Oakwood Boulevard off Michigan Avenue, the Ford engineering complex came into view. It was a long, two-story building with a facade of white granite. Henry Crecelius met Gregorie in the lobby and took him to the southeast corner of the engineering laboratory to show him where he would work. Later that day, Crecelius introduced Gregorie to a short, dark-haired man who was Edsel Ford. "I didn't make the connection with the gypsy's prediction at the time," recalls Gregorie. "It wasn't until years later that her words came back to me."

Crecelius, a soft-spoken man who spent most of his mornings at the Lincoln plant, put Gregorie to work drawing sketches for Lincoln bodies. There were only a few men in the whole department: Crecelius himself, a couple of body draftsmen, several detailers, and Gregorie, who was the only designer. The draftsmen had large drafting boards on which they worked, but Gregorie was given only a small table on which to make his artful renderings.

"The one thing I remember most about this time," recalls Gregorie, "was the 'foreign' cars lined up in the department. When I say 'foreign' cars, I don't mean cars from England or France, but non-Ford products, products in direct competition with Lincoln at the time: Packards, Cadillacs, and so on. Most of them were partially disassembled; engines were taken out and placed on dynamometers to check their horsepower and to test their durability. I think all that was Edsel's doing. The old man [Henry Ford] never cared much about what the competition was up to. But Edsel was very keen on that."

At the time, Lincoln, under the direction of Edsel, was producing two different chassis: a 136-inch-wheelbase chassis for production Lincolns, and a 145-inch-wheelbase chassis for custom bodied Lincolns. When Gregorie started working for Lincoln, Crecelius and his men were completing the design of bodies which the Lincoln plant would build in large quantities for the short-wheelbase chassis.

Many Lincolns at that time were touring cars with large fabric tops. They appeared sportier with their tops up than they did with their tops folded down, because when retracted the top became an unsightly pile of fabric extending past the rear of the body. This was contrary to Gregorie's sense of style, so he designed an ingenious top mechanism with moveable top bow pivot points. As the top was folded down, the bows moved forward on a plate attached to the body sill, thereby providing more area to store the fabric and minimizing the top overhang. With this idea, for which Gregorie received a patent, the touring car appeared equally sporty with the top down or up. More importantly, it was an early indication of how well Gregorie could meld mechanical aspects with design to improve the appearance of an automobile.

"I felt great about that," Gregorie recalls. "But that was one of the few contributions I made to the Lincoln line at that time. The whole body line was pretty well established when I started. There was no opportunity for me to do any real body designing other than making a change in a reveal molding or something like that."

Joe Galamb recalled that in those days Gregorie "was a very smart, cocky young fellow, and he wouldn't take anything from anybody." Normally, Edsel shied away from such overbearing people, but he and Gregorie seemed to get along well enough during their first few meetings that Edsel began to think Gregorie might be the one who could head a design department within the company.

CHAPTER THREE

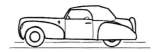

EDSEL FORD
Styling Executive

On August 28, 1957, among much fanfare and publicity, Ford Motor Company launched its first new model since the Gregorie-designed Mercury almost two decades earlier. Designed by one of the new breed of designers recruited by the company after Gregorie had left, the new model was glitzy, gaudy, and over-dramatic—the antithesis of the designs Gregorie and Edsel loved. It was long, loaded with chrome and gadgets, and had a very unusual oval-shaped front end treatment. Trying to be kind, some critics referred to the unique front end as a horse collar or a toilet seat when they first saw the new car. More critical observers said it resembled "an Olds sucking on a lemon" or "the female sexual organ."

The new model was designed at the zenith of albatrosses coming from Detroit during the 1950s, and the Ford Motor Company spent months trying to find an appropriate name for its new entry. After the marketing staff of the company failed to provide a suitable name, even after compiling a list of six thousand from which the executives could choose, it asked for assistance from the renowned American poet Marianne Moore. Her suggestions—Resilient Bullet, Arcenciel, and Aeroterra—were, to say the least, creative but unacceptable for the name of an automobile, no matter how bizarre the vehicle appeared. "May I submit," she wrote the company in one last attempt, "Utopian Turtletop"?

After months of wrangling, Ernest Breech, executive vice president of Ford at that time, settled on the one name that had not been on the exhaustive list but had been bantered around informally since the idea for a new line was first suggested, and he submitted it to the Ford family for final approval. The name he wanted was *Edsel*, named after Henry Ford's only child. Breech felt that this name would be a fitting tribute to the man who had been president of Ford Motor Company for twenty-five years and who was instrumental in dragging the company into the modern era, particularly through styling.

Breech anticipated a warm reaction from the Ford family, but instead he received only sighs of disbelief. Edsel's youngest son, William, who took after his father in appearance and love of automobiles, was adamantly against it. William said that other cars named after individuals, such as the Henry J, had a poor record of success, and he did not want a similar fate to befall a car named after his father. Benson Ford's reaction was, "Over my dead body." When a Ford representative went to see Edsel's widow, Eleanor Ford, to inform her of the decision, she slammed the door in his face. Henry Ford II did not think much of the choice either but reluctantly agreed to it to break up the log jam that had developed over naming the new car. It was a decision he would regret for the remainder of his life.

The Edsel entered the marketplace with a fizzle. Sales during the first few months were well below the break-even point, and they declined from there. Aside from its grotesque styling, the Edsel was rushed into production in an economy that had started falling into recession two years earlier. Quality problems soon began to surface, which caused the few buyers in the market to avoid the new model. By that time, the economy started to rebound in 1961, but Edsel sales had been so dismal during the preceding three years that the company decided to drop the line entirely in November 1959.

In the end, Ford Motor Company lost $350 million in its Edsel fiasco, and the name Edsel has been associated with disaster since that time. The humiliation was not so much for the monetary losses that the Edsel sustained, but because it tarnished forever the name of one of the most beloved yet obscure executives of the Ford Motor Company. "You're kind of sensitive," bemoaned William Ford, "when your father's name becomes a synonym for failure."

E.T. Gregorie had been long gone from the company when he heard the news that Ford was going to produce a new car and name it after his patron. "It was a travesty on Edsel's name!" Gregorie exclaims. "The name Edsel is not a particularly eloquent name, and it certainly wasn't appropriate for the name of an automobile. Maybe the company officials wanted to recognize Edsel's contributions, but the public did not make the connection."

The sorrowful story of Edsel the car was an ironic synopsis of the life of Edsel the man. (Figures 3-1 and 3-2) Born on November 6, 1893, Edsel Ford grew up to be a handsome man, with the finely sculpted facial features and thin build of his father and the dark complexion of his mother. He was well educated, a voracious reader, and had a "keen mentality," as his secretary once described. However, Edsel was a timid soul, not as gregarious and outspoken as his father, who thrived on being the center of attention. "Father likes that kind of thing," Edsel once said about his father's penchant for publicity. "I don't." In fact, Edsel made it a point to deliberately avoid the limelight, and he felt victorious when he could drive downtown without being recognized. Edsel did not like having his picture taken, even for company purposes, and although he was president of the company for a quarter of a century, Henry's picture hung in Ford dealerships instead of Edsel's. Edsel was so successful in maintaining his anonymity that many Ford dealers would not know him if he walked into their showrooms.

Henry appointed Edsel, then only twenty-five years old, president of the company on December 31, 1918. It was probably the biggest news event in the automotive industry since Henry had introduced mass production, but Edsel assumed his new position without fanfare. When asked to comment on his new appointment, Edsel replied humbly, using a Model T metaphor. "The hard part is finished," Edsel said. "It is easier to run something that has been cranked up and started and is running in high gear."

Away from the limelight, Edsel was enthusiastic about his new position and entered it full of gusto and with an air of confidence never evidenced by him in the past. "The Ford interests have always been a one-man proposition and always will be," Edsel proclaimed in the summer of 1924. "That's why Father stepped out and let me steer. I'm playing his game—doing things myself when I want them done right." It appeared as if Henry fully supported Edsel, allowing him to assume increasing responsibility. Instead of making an immediate decision when confronted by

Figure 3-1. Although this photo of seven-year-old Edsel Ford was staged for the camera, it is a fitting and accurate setting, for Edsel was a voracious reader. (Photo from the Collections of Henry Ford Museum & Greenfield Village)

Figure 3-2. This photograph of Edsel B. Ford was taken around the time the Continental was developed, circa 1939. (Photo from the Collections of Henry Ford Museum & Greenfield Village)

one of his subordinates as he used to do, Henry redirected them to his son. "I want you to take everything up with Edsel," Henry told them. "Things that you have been talking over with me, I want you to take up with him. Edsel is the boss."

Maybe Edsel believed that, and maybe Henry wanted to believe that, but all the men close to Henry Ford were dubious. "The smart ones," explained Fred Black, advertising manager at Ford, "also took up matters with the old man, usually saying, 'Well, I discussed this with Edsel, and he thought it was a good idea, but I wanted to be sure that you understood it and also thought it was a good idea.' "

The other executives were skeptical of Edsel's new power because they all had seen good men—talented men—come into the company and establish a foothold, only to be summarily dismissed, not because they were doing a poor job, but because Henry thought they were doing too good of a job and were gaining too much power. The executives who had survived this long had reason to believe that a similar fate would befall Edsel, son or no son of Henry Ford.

Despite their behavior, most Ford employees wanted Edsel to succeed. To them, he was the savior who would move the company into the next stage of its development. The manufacturing and production techniques so greatly influenced by Henry Ford were rapidly being replaced with administrative techniques, design, and sales. The employees believed that Edsel, with his education and experience, and the fact that he was part of a new generation receptive to these new developments, could move Ford in this new direction. Edsel, too, felt that his father was losing touch with the business, saying that his father "was more or less out of date."

The year 1926 would have been the perfect time for Henry to relinquish control of the company to Edsel. Henry was now sixty-three years old and deserving of a less grueling lifestyle. Edsel, now thirty-three years of age, was ready to assume control. Thanks to Henry, no other man in the company was probably more qualified. Even Henry Ford's long-time right-hand man, Charley Sorensen, could not match Edsel's skills in finance, administration, sales, and styling. Sorensen was an expert production man, but that was all he knew.

Unfortunately for Edsel and for the company, Henry never completed the baton pass to Edsel. The old timers were right. Henry never would give up control of his company—not completely. During Edsel's formative years, 1919 to 1926, Henry watched closely how Edsel managed and made decisions, and he was disappointed in what he saw. (Figure 3-3) What Henry wanted to see in Edsel was someone similar to himself—ruthless, cunning, and dictatorial. What Henry saw instead was someone who was considerate, unpretentious, and democratic—someone whom Henry considered "too soft" and "too easily influenced." As long as Henry felt that way about his son, he would never relinquish full control to him.

Unlike Henry, who was autocratic and often dictatorial in running the company, Edsel used a democratic style of managing, and he delegated responsibility whenever he could. He rarely gave orders, preferring instead to solicit input, suggestions, and recommendations from his managers and let them make their own decisions. Instead of saying "Do this" or "Do that," Edsel would elicit agreement and direction by offering suggestions. He approached problems analytically and impartially, and he relied on consultation and discussion with his subordinates to reach decisions. Edsel would not compromise between what he thought was right or wrong, but "he would seek adjusted agreement between extremes." Through his enlightened managerial methods, Edsel had as much control of the company as his father had and the respect of everybody, and he got things done through them. "The whole company," said one long-time Ford employee, "would go to blazes for him."

On the other hand, the elder Ford was guided by hunches and intuition, and he did not seem to care how anyone else felt. Henry intuitively felt he knew all the answers, and that "unless someone agreed with his position in its entirety, they did not belong in the company." "If I take my hunch," Henry explained to one of Edsel's schoolteachers, "I'm much more apt to be right than if I try to reason it out." In contrast, "Edsel had a reasoning mind and liked to sit down and talk things over" before forming a conclusion.

Edsel's management style befit his docile personality, and he had learned it from his one-time high-school manual arts instructor, Clarence Avery. Edsel was so impressed with Avery's mechanical horse-sense that he persuaded Avery to come to work for the company in 1912. Avery had a knack for solving complex problems—he was instrumental in helping Henry Ford develop the moving assembly line. However, equally important, Avery became Edsel's tutor in the art of managing. He taught Edsel that he could get the most from his men by leading them rather than driving them—a concept not only new to this cutthroat business, but one that was anathema to Henry Ford.

Unfortunately for Edsel and the company, Henry Ford did not understand his son's approach. He mistook it as a sign of weakness, and anybody who used it as inadequate. From Henry's perspective, anybody who adopted Avery's democratic approach to managing could not effectively control or direct anything, let alone the mammoth Ford industries.

Henry Ford eventually forced Avery from the company, as he would so many other Edsel advocates, and he began criticizing and ridiculing Edsel for accepting his old schoolteacher's methods. Although it is true that Henry needed others' assistance, he always made the final decisions. Being the boss and also asking for opinions from subordinates was contradictory to Henry, and he tried to teach Edsel that same reasoning. He felt that Edsel was listening too much to others' opinions, and he told Edsel many times, "Don't be swayed by what somebody tells you."

On this one point, Henry became obsessed. He wanted Edsel to change, and he began taking steps to force him to "be harsher." He thought his son needed to become "a little more shrewd, a

Figure 3-3. Edsel Ford (left) with his father (right), shortly before Henry appointed Edsel president of Ford Motor Company on December 31, 1918. Although only twenty-five years old at the time, Edsel went on to head the gigantic automobile company for almost a quarter of a century. Astute in all aspects of the company, Edsel's greatest contribution was in automobile styling. "In that," Henry Ford said, "I have a son that I can be proud of." (Photo from the Collections of Henry Ford Museum & Greenfield Village)

little more crafty, and…a little more tricky," in the same ways that he was. How Henry attempted to do that was the most tragic part of Edsel's life.

Many incidences over the years caused Henry to believe that Edsel was too easily influenced, but none were more poignant than Edsel's decisions to add more office space to the Highland Park administration building and to add more capacity to the steel-making operations of the company. In both instances, Henry abruptly and ruthlessly countermanded Edsel's decisions, hoping that this shock therapy would toughen Edsel.

During the halcyon days of the Model T, Edsel's subordinates, William Knudsen and Frank Klingensmith, convinced him of the need for additional office space and recommended that a wing be added to the existing Highland Park administration building. In his typical fashion, Edsel listened to their proposal, concurred, and ordered the construction to begin as soon as practical. Bulldozers arrived in short order and began digging a hole for the foundation, piling the dug earth around the hole. The contractor was making good headway until Henry came into Edsel's office.

"What's the construction for?" Henry asked his son.

"We need more office space," Edsel explained.

"What makes you say that?" Henry asked.

"Well, let's take a look, and I'll show you," Edsel replied, confident of his position.

As the two Fords walked through the administration building, Edsel showed his father the overcrowded offices and how additional desks lined the aisles similarly to cots in a hospital. After they had finished their tour, it was evident that more office space was needed; however, Henry outlined his position in short order, and it did not jibe with Edsel's viewpoint. "Edsel," he said, "we don't need that extra building."

"Well, all right, Father," Edsel acquiesced. "We'll close up the hole."

"No, don't do that," Henry replied. "Just leave it that way for a while."

As a penance to teach his son not to listen to other people, Henry ordered that the hole be left open to remind Edsel of his apparent blunder. What a penance it was. Henry made sure the hole was left open not only for a couple days or a few weeks, but for several months. The frequent Michigan rains quickly filled it with water, and mud from the piled dirt washed onto the pristine lawn and across the sidewalk. Everybody who entered the administration building had to walk around the mess every day.

As for the congested administration building, Henry rectified that, too, in his own demented way. To make room in the overcrowded offices, Henry purged the administration offices of half of its personnel. The employees weren't told if they were laid off; they found out if they still had a job when they returned to work the next morning. If their desk was there, they had a job. If their desk was gone, so was their job! "Are you still at Ford's?" one employee asked another. "Well, I was when I left this afternoon!"

After the purge, Henry called Edsel and they took another walk through the office building. "See that?" Henry asked rhetorically. "We don't need any more room. We've got plenty of room, haven't we?"

"Yes, Father," Edsel abashedly replied.

On another occasion, the affront to Edsel was equally cutting and the expense equally high. William Mayo was chief engineer at the time, and he convinced Edsel that the company needed additional coke ovens to support the demand for steel requirements. Again, after careful consideration and evaluation of all the data, Edsel told Mayo to proceed with building the coke ovens.

When Henry heard about the new project, he could not see the need for more ovens but he did nothing to stop the project. Henry should have stopped to the project right there, but he did not. In what was to become one of Henry's most contemptuous examples of ruthlessness and insensitivity toward his son, he waited until the ovens were completely built before making his opinion known to Edsel. Henry even told his right-hand man, Harry Bennett, of his intentions.[a] "Harry," he said, "as soon as Edsel gets those ovens built, I'm going to tear them down."

Whether that statement was made in confidence to Bennett is not known, but Bennett, in one of his more benevolent gestures (of which there were precious few for a man in his position), decided to warn Edsel of his father's intentions. However, Edsel found Bennett's story too fantastic. Why shouldn't he? Edsel knew Bennett was out only for Bennett, and this story was simply another example of Bennett trying to undermine him. By this time, who could Edsel trust?

"Bill knows more about coke ovens than Father," Edsel explained to Bennett. "I don't think he'll do anything of the kind." Nothing more was said, and Bennett waited for the explosion.

As Bennett had predicted, Henry Ford followed through with his threat when the coke ovens were completed. First, Henry fired Mayo. Then, as a climatic insult to Edsel, Henry ordered Mayo's son, who also worked at the company, to tear down the ovens. When Edsel heard what had happened, he shook his head and said, "I don't know what kick Father gets out of humiliating me this way."

We can argue about whether or not the company needed more office space or more coke ovens, but we can be sure that Edsel did the analysis to convince himself that it did. In any event, these two decisions prove Edsel's decisiveness more than his gullibility. Edsel may have been reserved and "soft-spoken," but he was not a pushover. He was "inclined to make a detailed study…gather all the facts," instead of making the snap decisions that his father was notorious for making.

This difference in approach to managing and decision making was the underlying reason for the derision that eventually developed between Edsel and his father. Neither Edsel's approach nor Henry's approach was inherently right or wrong. However, because Henry had made a number of uncanny business decisions in his career—the low-priced car, mass production, and increasing the wages of the laborer—all without making a detailed study of any of them, he thought his method was superior. He seemed to have forgotten that he had also made a number of out-of-hand decisions that were as devastating to the company as his propitious decisions were beneficial. Not replacing the Model T sooner (Edsel's idea), not changing from a four-cylinder engine to a six-cylinder engine (also Edsel's idea), and not incorporating hydraulic brakes sooner (another one of Edsel's ideas) were among Henry's mistakes. One engineer who worked for Ford for many years confessed that Henry Ford made him work on many projects that were totally unfeasible from the outset.

However, Henry's worst mistake was accusing his son of not having "enough boldness." Henry actually told one employee that he was "very disappointed with Edsel" and confided to Bennett that he thought "Edsel was a weakling."

What Henry never understood or refused to accept was that he and his son were complete opposites. They were as different as a Model T and a Lincoln. Henry once confided to his niece, Catherine Ruddiman, that "he and Edsel had not always understood each other, and at times could not see eye to eye." The way Catherine described the situation was that Henry and Edsel "did not speak the same language." In the same way as the homely Model T could go places the luxurious Lincoln could not, and the Lincoln was more suited to a night on the town than the Model T, the differences between Edsel and his father should have augmented each

---

[a] Harry Bennett was a hellion who joined the company in 1916. Henry Ford liked his daring personality, and over the years Henry gave him more and more power within the company. By the early 1940s, Bennett was director of personnel and Henry's close confidant. There was talk that Henry Ford admired Bennett more than his own son Edsel.

other, making each stronger and more resourceful. Instead, Henry allowed their philosophical and attitudinal differences to create an environment of conflict, driving each other apart, and causing Edsel untold frustration, humiliation, and sickness in the process.

Edsel worked under these hostile conditions throughout all his adult life. He could have called it quits a long time ago and lived the life of luxury. However, he was too loyal to his father to quit or to take a forceful stand against him. He always said that it was his father's company and that he would not interfere with the way Henry wanted it managed until his father was no longer in the picture. However, that never happened.

Fortunately for Edsel, one area of the company in which Henry had no interest was styling. Henry's life centered around production and the mechanical aspects of the automobiles he built, particularly the chassis and the engine. This left a void in the area of body design, which Edsel quickly and efficiently filled. Although the two men could have worked together more harmoniously, blending mechanical design with body design, Henry left Edsel alone enough in developing body designs so that Edsel was able to make Ford automobiles handsome in spite of his father's antiquated mechanical bents.

Edsel possessed artistic abilities and a natural affinity for design from an early age. Initially, he made doodles and incomprehensible sketches as most children do. As Edsel grew older, his childish sketches assumed more form and sophistication. By the time he was in high school, he could deftly paint a landscape, sculpt a figurine, or sketch a portrait in pastels. (Figures 3-4 and 3-5) For a while as an adult, Edsel pursued his passion for painting by taking lessons from noted painters in the Detroit area. "Edsel loved to paint," recalled the company's resident artist, Irving Bacon. "Several times he expressed a desire for me to come to his home to paint with him."

When most of his time was consumed in running the Ford industries, Edsel spent what little free time he had studying art history. He also was a founding patron of the Museum of Modern Art in

Figure 3-4. Edsel's artistic ability is evident in this charcoal still life he drew when he was about fifteen years old. (Photo from the Collections of Henry Ford Museum & Greenfield Village)

New York and was the president of the Arts Commission of the Detroit Institute of Art. (Edsel held that position for a quarter century.) Edsel's wife, Eleanor, loved art as much as he did, and they filled their beautiful Gaukler Pointe mansion with works of art by Cezanne, van Gogh, Matisse, and other masters. Nonetheless, Edsel's admiration of the great painters was surpassed by his passion for automobile design.

As early as six years of age, Edsel was drawing pictures of motorcars. (Figure 3-6) These pictures, as with his early drawings, were not pieces of art by any means, but they were remarkably well detailed. From wooden spoked wheels, brass headlights, brake levers, and chain sprockets, Edsel omitted no detail when he drew an automobile. In later years, his drawings were equally detailed but more refined. (Figure 3-7) They had better proportions, straighter lines, and symmetrical curves. It is unknown if Edsel

Figure 3-5. In this wallpaper design created when he was twelve years old, Edsel combines his penchant for art with his interest in automobiles. (Photo from the Collections of Henry Ford Museum & Greenfield Village)

 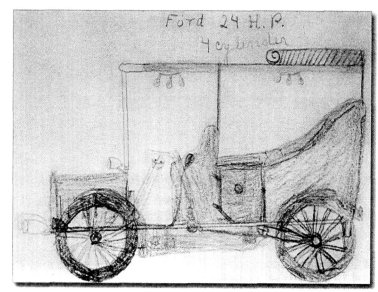

Figure 3-6. Although these early automobile sketches, done by Edsel Ford circa 1904 are quite rudimentary, they are well proportioned and detailed. (Photos from the Collections of Henry Ford Museum & Greenfield Village)

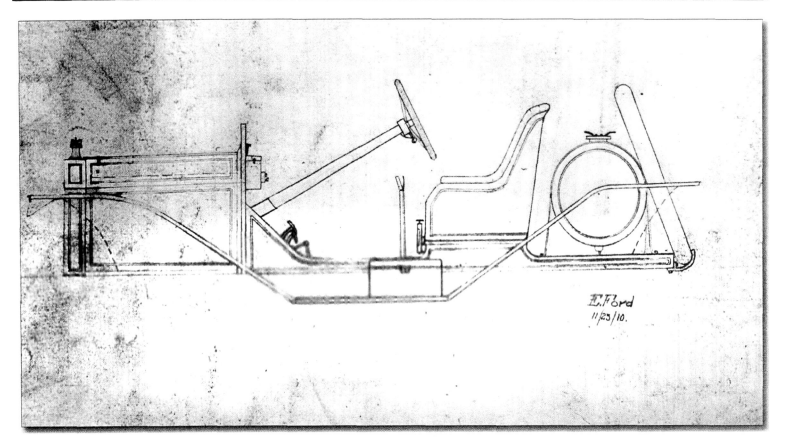

Figure 3-7. This expertly drawn speedster body was designed by a more mature seventeen-year-old Edsel Ford in 1910. (Photo from the Collections of Henry Ford Museum & Greenfield Village)

displayed any of his landscapes and still lifes in his boyhood home; however, Edsel's cousin, Catherine Ruddiman, recalled seeing pictures of automobiles which he had drawn in his high-school mechanical arts class, hanging on his bedroom walls.

Edsel further developed his artistic eye for automobile design by cutting pictures of cars from catalogs and magazine advertisements, pasting them in scrapbooks, and spending hours absorbing the appearance, construction, and personality of each vehicle.

He would not merely gloss over each picture; rather, he would analyze each automobile in detail, studying the contour of the fenders, the shape of the rear end, and the rake of the windshield, asking himself what made this particular car appear racy and that car appear elegant. Eventually, Edsel developed a sense of what looked good and what did not. This talent, more than his drawing ability or painting skills, proved beneficial to Edsel while directing E.T. Gregorie in designing Ford products throughout the 1930s and 1940s.

Henry promoted Edsel's interest in automobiles by providing him with an endless array of motorcars, much to the chagrin of Edsel's mother. Unlike Clara, who kept a watchful and discerning eye over her only child, "Father allowed me to do as I pleased," Edsel recalled. What Edsel pleased to do was "buy every new kind of motorcar that came out." Owning automobiles—all types of automobiles, especially exotic ones such as Packards, Locomobiles, and Rolls-Royces—became one of Edsel's passions. (Interestingly, he did not collect them, nor did he own them for very long. He simply bought them, drove them for a short while, and then sold them.) It was an expensive obsession, but with the phenomenal success of the Model T, the Fords had amassed enough wealth to afford Edsel this one indulgence.

Edsel's first car was a new, bright red 1903 Model A Ford Runabout, the first model manufactured by his father's newly established Ford Motor Company. Edsel was only ten years old at the time, and he was much too young and slight of build to crank the engine himself. However, Henry taught Edsel how to drive and operate the car with every competence other than that, and Henry instructed his butler to crank the car for Edsel if Henry was not there. Because there were no driving restrictions in those days (the authorities thought that if you were old enough to drive a horse and buggy, you were old enough to drive a buggy without a horse), Edsel drove his red runabout to school, to the store, and all over town without reservation. This made Edsel an instant celebrity with all the children in the neighborhood. He gave them "joy rides" in the summer, and in winter he pulled them on sleighs—three and four at a time—behind his little Ford along Hendrie Avenue, where he lived. (Figure 3-8)

Three years later, Edsel got his second car, a 1906 Model N Ford. A compact, highly reliable car that was a precursor to Henry's famous Model T, the Model N was powerful enough to break the arm of any man who cranked it improperly, and that concerned Henry. "If you ever go downtown and can't get it started," Henry instructed, "call your mother and she'll get in touch with the butler."

"As I was able to handle them and knew everything about them in the way of performance," Edsel explained about his ever-changing array of motorcars, "my father provided me with one [car] after another." Before long, Edsel had acquired a stable of motorcars, most of which were not Fords. (Edsel always felt more comfortable driving a Locomobile, Marmon, or Lincoln than one of his father's Flivvers.) At one accounting, Edsel had twenty-seven cars in his stable, ranging from a Ford stake bed truck (presumably for the hired help) to a Lincoln Town Sedan and a Bugatti sports car.

In his late teens, Edsel began incorporating the things he had learned about automobile design into specially built motorcars. Most of his early one-of-a-kind automobiles were merely revamped Model T's with interesting design cues not available on regular production Fords, such as monocle windshields, dual bucket seats, exposed gas tanks, swooping fenders, and sporty wire wheels. (Figure 3-9) We get a rare glimpse into the intensity Edsel placed in these projects, as he did in most of his endeavors, from a notebook he kept. In it, he listed the things he had to do day by day to complete one of his speedsters: "separate engine from muffler..., lengthen steering post..., turn over steering wheel..., draw hub caps..., make bearing retainer..., shorten running boards..., install rubber fenders..., remove 2 leaves out of rear spring..., color: gray, white pin stripe."

These modifications were relatively easy to incorporate into a production automobile, but Edsel went further later. In one car, he had Ford engineers design and build a pair of bucket seats, a set of wire wheels, a V-shaped radiator, and a complete six-cylinder engine to place in one of his speedsters. (Figure 3-10) "Mr. Edsel Ford often came in [the shop] and watched us work on it," recalls one of the engineers assigned to the project. "He was a quiet man, very retiring, but he had a real smile on his face as he watched us work on...his car."

Compared to regular Model T's (which had square, top-heavy bodies mounted on wooden-spoked wheels), Edsel's personally designed speedsters appeared quite exotic and racy. This

Figure 3-8. Ten-year-old Edsel was the hit of the neighborhood in the winter of 1906. Here he pulls a group of neighborhood children on sleighs behind his new Model N Ford. (Photo from the Collections of Henry Ford Museum & Greenfield Village)

Figure 3-9. Edsel Ford is behind the wheel of one the Model T speedsters that he designed and built himself, circa 1910. Cars such as this helped Edsel develop his eye for automobile design. (Photo from the Collections of Henry Ford Museum & Greenfield Village)

Figure 3-10. Edsel Ford sits behind the wheel of his six-cylinder speedster, circa 1912. Most of the parts of this car, including the engine, were specially made for Edsel by Ford engineers. (Photo from the Collections of Henry Ford Museum & Greenfield Village)

most recent car, as one of the engineers explained, "was the latest word in speedster."

From these early design experiments, Edsel applied his design talents to a production vehicle for the first time in 1910, when he was only seventeen years old. Taking a run-of-the-mill Model T runabout, Edsel stripped its body down to the frame and rebuilt it into a sportier motorcar. He lowered the seats and the steering column, added a slightly longer hood, and added doors to the open body. Taking cues from his speedsters, Edsel mounted a cylindrical gas tank directly behind the seats. Although the changes were minor, the results were astonishing. Edsel's mass-produced Torpedo was a Model T, but it looked faster and more refined than his father's standard runabout. (Figure 3-11) More importantly, Edsel showed that it was as easy to manufacture a well-designed automobile as it was to make a strictly functional one.

From a design perspective, Edsel's Torpedo was a success, but Henry did not take Edsel's initiative well. As Edsel quickly learned, it would take more than one design triumph to convince his father to change a square-shaped fender to a rounded form, or to persuade him that an enclosed car did not have to be tall enough to accept a gentleman and his top hat at the same time! "Henry was dead set on the idea that a utility car did not have to have style and comfort," says Gregorie, and Henry did not try to hide his partiality. "Is it not better to sacrifice the artistic to the utilitarian than the utilitarian to the artistic?" Henry asked rhetorically. "What, for instance, would be the use of a teapot that would not pour because of the ornate design of its spout?" That is the mindset Edsel Ford had to deal with.

Another reason Edsel had difficulty in persuading his father to change the Model T was that the car was such an astounding success as it was. Based on Henry's philosophy of low cost and high utility, his homely creation created a demand that exceeded its supply from almost the moment it was introduced. Sales of the car grew at an average annual rate of more than fifty percent over

Figure 3-11. The 1912 Model T Torpedo roadster (top), designed by Edsel Ford, has a more complete and streamlined appearance than his father's standard Model T roadster (bottom). (Photos from the Collections of Henry Ford Museum & Greenfield Village)

the first ten years of its life. Such spectacular sales kept Henry's production lines humming—as long as they hummed, Henry was content.

However, the tremendous success of the Model T camouflaged the prodigious sales that the remainder of the industry was experiencing, especially those of Chevrolet. Introduced in 1912, a few years after the Model T, but directed one notch above the same market, the Chevrolet was enormously popular. From 1925 to 1926, its sales rose sixty percent. As the 1920s came to a close, Chevrolet was already selling at half the rate of Ford—almost 700,000 vehicles a year, compared to Ford's 1.4 million vehicles.

Edsel was uncomfortably aware of the pending threat posed by Chevrolet, and he knew the General Motors strategy all too well: offer more style at a slightly higher price, and you could combat the mighty Ford Motor Company which was, at that time, the largest automobile manufacturer in the world. Europeans had known the importance of styling a long time before Chevrolet discovered it, and so did Edsel. For instance, Edsel did not have to be unsympathetically reminded by an English visitor about the mood of the marketplace, "You know, we like to sit *in* our cars, not *on* them."

Concerned with the inroads Chevrolet was making, but hobbled by his father's one-track mind, Edsel realized that it would take time and patience to convince Henry to make any major changes on the Model T. And asking for a totally new car? Well, that was simply out of the question. Nonetheless, Edsel knew he had to try, so he took a gentle and persuasive tack with this father, hoping to persuade Henry to make modest changes to the Model T over time and thereby easing Henry toward the inevitable—replacing the Model T with something more elegant and fashionable as well as utilitarian.

Edsel's tenacity and patience were successful, at least to a certain degree. In the fall of 1922, more than a decade after his first successful production design, Edsel received a belated opportunity from his father to make some stylistic changes to the Model T. Henry told Edsel that he could make any cosmetic changes he wanted, but he was to leave the mechanics of the car alone—a philosophy Henry maintained until his death. Edsel was not fond of mechanics anyway, and because he did not receive approval to make changes until late in 1922, he would not have time to make any substantive mechanical changes. He barely had time to make stylistic changes in the Model T.

Nevertheless, with the help of Joe Galamb, Ford's chief body engineer, Edsel managed to dress up "Lizzie" by incorporating more curved surfaces and smoother lines in the body and fenders, as he had always wanted. He also added height to the radiator, blending its decorative apron at the bottom into both fenders, and he raised the lines of the hood so that it flowed more graciously into the cowl. These were simple changes, but they gave the car a more solid appearance and the impression that it was designed with an overall objective in mind.

Although these changes were by no means everything that Edsel might have wanted to do to the stodgy Ford, they were nonetheless beneficial, exemplifying the power that styling was already achieving in the industry. Although the changes in 1923 were minor, Model T sales reached their zenith that year, exceeding two million vehicles.

If such subtle changes as were made to the 1923 Ford could cause such a boon in sales, reasoned Edsel, then making substantial modifications should produce even greater rewards. While his father was away in Europe, visiting his foreign factories, Edsel decided to press forward on a design prototype the Ford employees called the Australian Job. (Figure 3-12) Starting with a standard Model T touring car, Edsel instructed his engineers to cut, chop, section, and weld the body and frame of the car until they made a car that was lower and longer. By the time they were finished, the engineers had removed no less than four and one-half inches from the ground-to-roof height of the car, and they added twelve inches to its overall length, dramatically softening the one-time top-heavy and stubby appearance of the Model T. The final touches included a one-piece windshield and a coat of bright red lacquer.

"It was a beauty," exclaimed George Brown, the purchasing agent who had bought many of the special parts for the car. "You put that in a showroom window, and you wouldn't need salesmen!" Edsel thought he had a winner, too, and he placed the prototype in the officers' garage in the Highland Park plant to await his father's return. He wanted to beat his father to the office so that he could be there to explain the purpose behind building the handsome touring car. To everyone's surprise, Henry decided to go directly to the plant the evening he returned to Detroit, instead of waiting until morning as everyone had anticipated. Henry's chauffeur dropped him off at the Woodward Avenue entrance, where he proceeded directly up to his office. From there he went downstairs to the garage, where he ran into Brown, who had been working late.

"I glanced up," Brown recalled, "and, oh, my God, my heart stopped! There was Mr. Ford at the top of the stairs, looking down at the car."

Figure 3-12. This experimental Ford design of August 15, 1924, was drawn by Gene Farkus. This blackboard drawing was made around the time of the Australian Job and could actually be a sedan version of the touring car that Henry Ford single-handedly demolished. This car is significantly longer and lower than its Model T counterpart, has room under the hood to accommodate a six-cylinder engine, and indicates Edsel's desire to add some class to his father's Tin Lizzie. (Photo from the Collections of Henry Ford Museum & Greenfield Village)

Brown hesitantly met Henry at the foot of the stairs. They stood there and made small talk for a few minutes. All the while, Henry was eyeing the bright red mock-up.

"What's over there?" Henry finally asked suspiciously.

Almost overcome with anxiety and hardly able to speak, Brown muttered, "Well, Mr. Ford, that's the new car."

Henry stared at it some more, then questioned, "Ford car?"

"Yes, sir," Brown replied.

"It is?" Henry responded in bewilderment.

Henry kept staring at the car as he walked toward it. "He'd tip his head this way, then he'd tip his head the other way," Brown explained. "He walked around the car three or four times, looking at it very closely. Finally, he got to the left side of the car, and he gets hold of the door, and BANG! one jerk, and he had it off the hinges! He ripped the door right off! God, how the man done it, I don't know! He jumped in, and BANG! goes the other door! BANG! goes the windshield! He jumped over the back seat and started pounding on the top. He wrecked the car as much as he could."

Brown ran up to the second floor, where some of the engineers who had helped develop the car were working. "Look out, boys!" he cried. "Mr. Ford's kicking the car all to pieces downstairs!" They all jumped up and ran down to the garage to see the commotion. When they arrived, Henry was still at it. "His hands were going, his feet were going, and talk about cussing! It was the first time I had ever heard Mr. Ford cuss."

It was a major setback for Edsel and indicated what he had to endure from his father. Edsel thought he had finally persuaded the old man to give at least as much credence to design as he did to engineering, and he wanted to prove what could be done to the stodgy Model T, given enough time and resources. However, whatever styling propensities Henry accepted with the changes Edsel made in 1922 were erased with the Australian Job. To Edsel's credit, however, he did not give up.

Ironically, if Henry had agreed to produce Edsel's Australian Job, perhaps he could have stemmed the precipitous decline in Model T sales that began shortly after the garage scene. Even with the changes incorporated the preceding year, Model T sales dropped in 1924 and again in 1925, and by 1926, they were plummeting. Edsel pressed his father for more styling changes in an effort to stem the regression. As far as Edsel was concerned, the only way to regain sales would be through additional aesthetic changes— or, better yet, by introducing an entirely new car. However, Edsel wanted more refinements than his father would ever think of giving, at least for the moment.

Meanwhile, Edsel had persuaded his father to purchase the bankrupt Lincoln Motor Company from its founders, Henry and Wilfred Leland. (Figure 3-13) In 1917, the Lelands established a manufacturing plant in Detroit to build Liberty engines—massive V-12 engines used in both airplanes and boats—for the military during World War I. When the war ended, they converted the plant into making luxury motorcars. They experienced a relatively good start, but an economic recession hit the country in the fall of 1920 and automobile sales dropped, particularly those of luxury motorcars, forcing the Lelands into receivership.

At first, Henry was not interested in absorbing a luxury line into the Ford fold, but he finally agreed to purchase the beleaguered company, ostensibly to help his old friends Henry and Wilfred. In fact, Henry saw it as a way to get Edsel off his back "and give the boy something to do," according to Charles Sorensen, production chief at Ford. That is only half the story. Actually, Edsel wanted the company as much as his father wanted him to have it. Although it added to Edsel's already burdensome load, he successfully turned Lincoln around and proved to everyone, including his father, that he knew how to run an automobile company. "Lincoln was Edsel's domain," Gregorie explains, "and

Figure 3-13. Edsel and Eleanor Ford pose with their 1922 Lincoln sedan parked in front of the Lincoln Motor Company administration building on February 4, 1922. Moments later, Edsel signed the agreement purchasing the company from its founder, Henry Leland. During the next decade, Edsel, working with numerous custom-body suppliers, would make the Lincoln line one of the most fashionable in the industry. (Photo from the Collections of Henry Ford Museum & Greenfield Village)

surprisingly, the old man let him run it pretty much the way he wanted to. In fact, if it weren't for Lincoln, Edsel and I would not have been able to come up with some of the more unique cars that we did."

The key to revitalizing Lincoln was styling. The quality of the Leland-Lincolns was unquestionable, but their styling was drab and homely vis-à-vis other luxury manufacturers. When Edsel took control of the company, he maintained the high quality of the Lincoln but markedly improved its styling by soliciting the talents of numerous custom coachbuilders located throughout the country. There was hardly a custom coachbuilder Edsel did not try over the years, but he tended to favor the designs of J.B. Judkins in Merrimack, Massachusetts; Willoughby & Company in Utica, New York; Brunn & Company in Buffalo; LeBaron, Carrossiers in Manhattan; and Dietrich, Inc. in Detroit.

Occasionally, Edsel ordered bodies from the coachbuilders' catalogs, but usually he worked directly with the carrossiers and their designers to develop designs to his liking. He frequently offered design ideas up front for the designers to consider, but, curiously, he never took a pencil in hand and marked a designer's drawing to indicate what changes he wanted made in the proposal. Nor did Edsel ever present a sketch of a design that he had made himself and say to the designer, "This is what I have in mind," although he had the artistic ability to make a design sketch as well as the designer himself. Instead, he preferred to explain verbally what he wanted changed and then allowed the designer to interpret his verbal directions on paper. The most Edsel would give to the designer in writing was a list showing the type of body he wanted for a particular job (such as a roadster, phaeton, or close-coupled sedan), the color scheme he preferred, and the kind of interior fabrics he favored. (Figures 3-14 and 3-15)

Figure 3-14. This was Edsel Ford's personal 1923 Lincoln coupe, body by Locke. Edsel approved the design of every Lincoln that came from the Lincoln Division between 1922 and 1943. (Photo from the Collections of Henry Ford Museum & Greenfield Village)

Figure 3-15. Most of Edsel's custom-bodied motorcars were Lincolns, but occasionally he would indulge himself with exotic European motorcars. Here he poses for the camera as he sits in a magnificent 1912 Rolls-Royce Silver Ghost touring car. He is only eighteen years old and had recently graduated from high school. (Photo from the Collections of Henry Ford Museum & Greenfield Village)

This was agonizingly detailed work, but no one enjoyed this protracted process of custom-body making more than Edsel. He followed every custom-made Lincoln body from inception to completion as if he were the owner of each one. His eye for detail and his exacting personality forced him to analyze every new design down to the most minute detail, in the same way as an accountant searching for a lost penny. Correspondence between Edsel and a chassis supplier illustrates his penchant for detail.

"We are wondering," wrote the president of Packard Motors, "if there is any particular reason for your desire [on the chassis you ordered] to have the extra tires rib tread? The standard rear tires on our cars are Goodyear cord, All Weather tread, and all our customers so far have desired this tread for their extra tires."

"I would say," Edsel replied, "that the specifications as made out are all right; that is, rib tread is desired, and I would not care for the All Weather."

Custom-body designer Raymond Dietrich soon learned of Edsel's resoluteness in a letter Edsel sent to him, complaining about Dietrich not following directions. "First of all," Edsel wrote, "I do not think that your production model is anything like the first sample that was built.... The detail is quite different on the interior and to my mind the job is a very inferior model and does not in any way come up to the standard that we set in the Lincoln body types."

Unquestionably, when Edsel gave a body designer specific instructions, he fully expected them to be carried out, exactly as he specified. He assured compliance by demanding that he approve every step of the design process. "I think you are working entirely too fast on this job," Edsel wrote to Hermann Brunn about a new Lincoln body Brunn was developing. "On the first sketch that was submitted, you had drawn up the working drawings before I had a chance to approve or disapprove it, and now you have the working drawings finished again.... This all costs money and seems foolish until I am satisfied with the sketch. I have not given you the go-ahead, and please do no more until I do so."

Only after spending countless hours specifying and approving every detail on every job would Edsel approve any of the carrossiers' design proposals. Such in-depth analysis may seem obsessive, but it was one of Edsel's greatest strengths. Through such painstaking examination—viewing and reviewing every line and contour of every design—he was able to identify the nuances that could change a mediocre design into a classic one. Once Edsel visualized the overall design or worked out a specific detail, his internal deliberations would end and he would pursue the design or detail doggedly to a conclusion.

"In those days," says Gregorie, "instead of taking the overall design of the car into account, it was customary for the chassis people to build the chassis and also develop the radiator grille, the hood, the front fenders, the placement of the head lamps, and the dash panel. The chassis, with all of these parts attached to it, is what the custom-body people received to put their bodies on. In other words, the Ford Motor Company never designed the Lincoln as a complete car—just the front end of it—until they started building in-house bodies at the Lincoln plant along about 1932. Before that, what Joe Galamb and the others did was prop up a grille on some sawhorses and hang a couple of fenders, and then stick a couple of dummy wheels under them to see how it looked. That's what Edsel looked at—without a car behind it—when making his front end selections. I used to kid him about that. I'd say, 'That's like going to a tailor to have your suit made, and he'd show you a couple of pants legs, and then he'd hold the coat up, instead of letting you try it on to see how the damn thing fit!' Oh, he'd get a big kick out of that! Actually, those Lincolns had some pretty nice-looking front ends when you consider how they were developed. Edsel had a lot to do with those coming out right."

Aside from catalogs and magazine advertisements, the best place to see exotic automobiles outfitted with the latest custom coachwork was at the Importers Salon held at the Commodore Hotel in New York City each fall. All the custom-body houses, both domestic and foreign, displayed their latest designs there. Accompanied by his Lincoln sales staff, by the renowned body designers from LeBaron, Carrossiers—Ray Dietrich, Tom Hibbard, and Ralph Roberts—and later by E.T. Gregorie, Edsel painstakingly evaluated every motorcar in the exhibit. Although dozens were on display, he did not miss any detail, no matter how trivial, that distinguished one elegant design from another. First, Edsel viewed each motorcar from a distance, as though he were pondering a fine Rembrandt or Monet. Then he would analyze it in more detail by walking up to it and inspecting its workmanship, feeling its upholstery, and climbing into the driver's seat to get the feel of the controls.

The Lelands had sold only slightly more than three thousand Lincolns during the seventeen months they built the car. Edsel almost doubled that amount during the first eleven months he controlled Lincoln and almost tripled production by 1927. Mechanically, Edsel was building the same car the Lelands had been building in 1920; however, stylistically, his cars were urbane, sleek, and elegant. That is what made them sell.

Faced with ever-decreasing Ford sales and realizing how his son had almost single-handedly turned around Lincoln through his styling efforts, Henry finally capitulated in August 1925, giving Edsel permission to make the most pronounced changes in the Model T since its introduction almost two decades earlier. Taking cues from what he had learned with his custom-bodied Lincolns, Edsel enhanced the curvature of the fenders and lowered the height of the Model T. Its chassis was lowered one and one-half inches, the radiator was raised, and a redesigned cowl merged with longer body lines. (Figure 3-16) In all models except the Fordor, the steering wheel was dropped three inches, and the seats were lowered and moved backward to increase leg room. In an effort to emulate the Chevrolet, Edsel

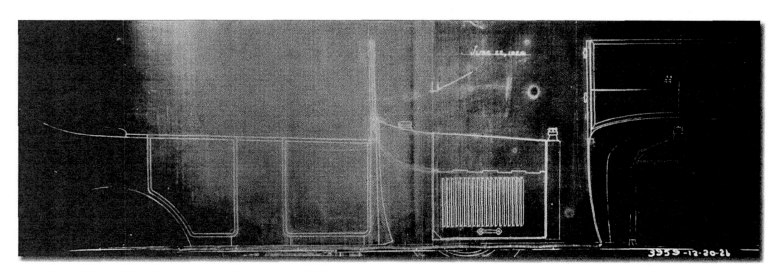

Figure 3-16. This blackboard drawing, completed in June 1926 by Gene Farkus, a Ford engineer, is an early design proposal for the new Model A Ford. The radiator shell is vintage Model T, but the remainder of the body design is similar to how the Model A would appear. (Photo from the Collections of Henry Ford Museum & Greenfield Village)

introduced new body colors on closed models, such as Channel Green, Windsor Maroon, Fawn Gray, and Gunmetal Blue. Wire wheels, made optional on the new design to give the car a sportier and more contemporary appearance, were available in Casino Red, English Vermilion, yellow, and black, offering contrast to the new body colors. It was exciting for Edsel finally to see a Model T outfitted with a green body and contrasting vermilion wheels after the millions of Lizzies that had been produced in Henry's ubiquitous black. Charles Sorensen, who had overseen the production of every one of the more than fifteen million Model T's produced, agreed. "I was sick of looking at them," he said.

Unfortunately, Edsel's changes were too few and too late. Although the new Model T appeared more stylish, it was nonetheless a Model T. Except for its outward appearance, everything under the skin—engine, drive train, suspension, and steering—was 1908 technology. In fact, with Edsel's new larger and heavier bodies, the Model T barely had enough power to get out of its own way. Wrote one Ford dealer about the so-called improvements, "Yes, you can paint a barn, but it will still be a barn...."

Edsel knew that a completely new automobile was needed to satisfy the public's growing demand for more refined cars. Finally, he told his father what he had known for years. "Father," insisted Edsel, "we've got to dress up the Ford. The Model T will have to be scrapped so that we can do something about the competition Chevrolet is giving us!"

In the fall of 1926, with Model T sales having fallen almost twenty-five percent since their peak three years earlier, Henry Ford finally decided to begin work on an entirely new automobile. Holding true to his character, he decided that a totally new model, not simply a redesign of his beloved Model T, would be developed. If the public wanted a new Ford, he said, a new Ford they would get. Except for the size of the tires and Ford script on the radiator shell, nothing on the new Ford that he and Edsel engineered and designed would either fit or resemble anything on the old Model T.

To Henry's credit, he finally acknowledged Edsel's grasp of industrial design and allowed Edsel a free hand in controlling the body styling of the new Ford car. For the first time, father and son worked together to develop a Ford. Edsel took care of the body design, while Henry directed the engineering. Henry had a complete engineering staff to assist him, but Edsel had only Joe Galamb. For additional design expertise, Edsel turned to the Murray Corporation of America and the artistry of Amos Northup. Edsel had already contacted Murray in April 1926, months before his father had given approval to build an entirely new automobile. (Figure 3-17) Using the same process with which he had built countless custom-bodied Lincolns, Edsel had sent to Murray a 1926 Model T chassis, complete with radiator shell, hood, front and rear fenders, running boards, and splash aprons. He had asked Murray to design and build a new body for it that was more "beautiful in appearance, [with] European lines, wider doors, and low effect." By November, Murray had a complete prototype ready for Edsel's inspection. (Figure 3-18) "The job...was constructed not to alter in any way the interchangeability feature of your chassis...which you laid down...as a fundamental," wrote a Murray executive. "We have, however, put on smaller wire wheels of 27 × 4.40 size. We used crimped springs...which brought us possibly one inch closer to the ground."

The design that Murray developed was a handsome car, even if it rested atop a Model T chassis. However, the design went no further than this one prototype. By the time the prototype was completed, Henry had given the okay to pursue a completely new automobile, and the initial Murray proposal was dropped. At that point, Edsel, Joe Galamb, and Amos Northup began work on a completely new body design based on an entirely new chassis. As with the earlier proposal, the new body was designed at a remote plant owned by Murray. "Joe Galamb and Edsel schemed up the old Model A mostly with blackboard drawings—chalk drawings on a blackboard," says Gregorie. "They never made clay models then or anything of that sort. You couldn't tell very much how the car would look until a prototype was made. It was all very crude and unprofessional."

Figure 3-17. This rare photograph shows a designer at Murray Corporation developing a clay model, circa 1937. The vehicle appears to be a convertible sedan and probably is being developed for Edsel Ford's consideration. (Photo from the Collections of Henry Ford Museum & Greenfield Village)

Figure 3-18. This prototype body was designed at Murray Corporation. At left is a production version 1927 Model T sedan, and at right is Murray's proposal. Although the Murray body is sitting on a production Model T chassis, as Edsel instructed, the car appears more stylish and modern than the moribund Model T. In fact, the prototype is taking on the lines of the soon-to-be-released Model A. (Photo from the Collections of Henry Ford Museum & Greenfield Village)

Figure 3-19. Early blackboard design of the Model A Ford, February 13, 1926. This side view is already taking on design cues of the production 1928 Model A: hood, cowl, and window shapes. However, the fenders still resemble those of a Model T. (Photo from the Collections of Henry Ford Museum & Greenfield Village)

Nevertheless, along with Joe Galamb, Edsel scrutinized every aspect of the new model as Amos Northup developed it. When different ideas were presented, Edsel and Galamb would debate them at length, in the same way as Edsel did in working with a designer on a custom Lincoln body. No detail was too minute or feature too trivial to warrant their scrutiny. They debated the profile, the ride, the materials used for making the seat cushions, and even what windshield to install to give good vision. "Edsel knew what he wanted," Galamb explained, "and insisted that he get it."

One of the details to which Edsel paid particular attention was the front end of the car. He knew from years of experience that the front end of the new Ford would receive the most critical attention; therefore, he insisted that it have an appropriately designed grille. Eugene Farkus, a long-time Ford engineer, was given the assignment.

"Initially, we did not have any grille on the front, just the radiator itself," Farkus explains. "Edsel wanted a grille, but he did not want it perfectly flat like the Model T's was. He wanted to have a little body to the shapes, so I bumped it out a little bit so that it had a curved, warped surface. In its final form, the grille had a little reverse peak in it—just for style." (Figures 3-19 and 3-20)

As Edsel had predicted, the grille of the new Ford received the most attention, but that attention was acclaim rather than criticism. Rounded at the top and squared at the bottom, the grille

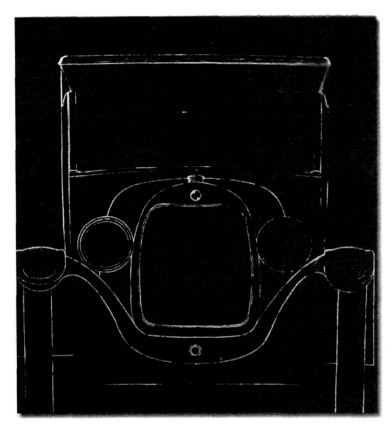

Figure 3-20. This front view of the 1926 blackboard drawing shows the classic radiator shell shape of the Model A, designed by Farkus with Edsel's direction. (Photo from the Collections of Henry Ford Museum & Greenfield Village)

was perfectly proportioned. With Farkus' reverse peak feature, it became the focal point of the entire car and enabled the rest of its design to have low contours.

To complement the striking appearance of the Model A, Edsel devised various color schemes for the new car. Unlike the last versions of the Model T, for which he added only a handful of colors, Edsel provided a complete palate of body colors for the new Model A. Henry Ford's favorite color, black, was a given, but Niagara Blue, Dawn Gray (light or dark), Arabian Sand, Copra Drab, and Andalusite Blue were among the other colors on which Edsel insisted. "The joke on me," Harley Earl lamented at the time, "is the painting of [the new Ford] is almost exactly like my La Salle, which I had painted as a special job. Hereafter, I shall have to go to Highland Park for my coach work!"

Introduced in the fall of 1927, the new Model A instantly boosted the prestige of the company, giving Edsel a great sense of gratification. (Figure 3-21) The Cartier brothers wrote to Edsel, applauding him on the design of the new car. "Perhaps you could let us mount a few Model A's up as broaches and bracelets," wrote Pierre Cartier. "I could personally sell fifty on Monday." The new Ford showed the public and the competition that Ford Motor Company was a major player in the industry and that Henry Ford was finally willing to give as much credence to design as he had to engineering. It was Edsel's greatest achievement to date. "The mere fact that Edsel was able to take a country-boy vehicle," explained one reporter, "and extract from it something beautiful, tractable, and urbane, testifies to his inner strength of purpose."

Henry Ford preferred the useful "Tin Lizzie" to its more graceful and comfortable successor, but realistically he understood that the success of the Model A was primarily a result of Edsel's design mastery. During a conversation with an old acquaintance, Henry credited his son with giving the new Ford its "style and much besides." From that point, noted Charley Sorensen, Henry would not permit anything to leave the Ford factories "until Edsel had passed on it." It was indeed a triumph for Edsel to have his father finally and sincerely recognize his talent. (Figure 3-22)

Raymond Dietrich, who worked with Edsel on numerous design projects, attributes Edsel's design talent and affinity for style to his "discerning eye." Edsel understood the difference between an "ordinary line" and the lines of the "classics," Dietrich explained.

Figure 3-21. Henry Ford (left) and Edsel Ford (right) at the time of the introduction of the Model A in the fall of 1927. For years, Henry had stymied Edsel's desire to add class to the utilitarian Ford. It was not until Model T sales declined precipitously that Henry turned over the body design of Ford automobiles to Edsel. (Photo from the Collections of Henry Ford Museum & Greenfield Village)

Figure 3-22. Edsel Ford drives a new Model A Ford roadster from Ford's engineering laboratory in Dearborn. More than anyone, Edsel was responsible for the styling of this beautiful and successful line of Ford cars. (Photo from the Collections of Henry Ford Museum & Greenfield Village)

He had an "instinctive liking for dignity in an automobile—dignity that reflected its purpose." To Edsel, every line, contour, and angle in an automobile design must have intent and simplicity. Nothing could be extraneous. Gregorie agrees. "If Edsel's taste in design could be summed up in one word," he explains, "it would be *finesse*. He had a very strong feeling of the pure sensitivity of a line and a feeling of lightness." Big, bulbous designs with thick, heavy bumpers and the excessive use of chrome on an automobile were anathema to Edsel. "He wanted delicate bumpers, delicate door handles, and things like that," says Gregorie. For these reasons, the automobiles developed under Edsel's tutelage had tight, smooth, conforming skin, and a trim, delicate effect. Unfortunately, Edsel's preference for fineness caused Henry to think that his son was "too artistic" for the automobile business, "if I can put it that way," says Gregorie.

To a great extent, Edsel's artistic bent was influenced by his love of boats and conservative nature. "He had his own way of determining whether something met with his taste," says Gregorie, "and his taste was very conservative. I don't know if he just did not care for anything spectacular, or just did not dare stick his neck out toward anything that was spectacular. That was just his nature. I believed he always felt on safe ground by following a good middle ground."

In 1935, Edsel made the list of best-dressed men, along with actor Fred Astaire. "Whatever he wore, owned, or drove had to be severely simple and plain, but in good taste," says Gregorie, "*immaculately* good taste." A *Fortune* reporter wrote in 1933, "All the Ford executives dress neatly, conservatively, but Edsel is the only one you may notice is wearing the green-and-white tie today, was wearing the brown-and-white tie yesterday."

Edsel owned many graceful yachts and sleek speedboats throughout his life. In summer he could be found off the rocky coast of Maine sailing his 63-foot *Acadia*, (Figure 3-23) and in winter his beloved *Onika* along the sun-drenched beaches of south Florida. During the 1920s, he participated in the Gold Cup Regatta on the Detroit River in his 30-foot speedboat, *Goldfish*. (Figure 3-24) In Edsel's mind, the sharp angles and uncomplicated lines of his yachts and speedboats were the epitome of gracefulness and simplicity, and he tried to carry those themes into the automobiles produced by Ford.

From the early Lincolns to the Model A and all through the Fords of the 1930s and 1940s, Edsel personally selected all the exterior colors and interior fabrics. Salesmen from paint suppliers and fabric suppliers would deluge him with new color schemes and fabrics every year, but Edsel always settled on earth tones and muted colors such as Coach Maroon, Dartmouth Green, and Folkstone Gray for paint, and taupe broadcloth and taupe mohair for interior fabrics. In fact, Edsel's favorite color was gray, and he had most of his special cars painted in gray and upholstered in gray leather.

For all his artistic ability, his involvement with the design of hundreds of custom-bodied Lincolns, and his direction of the design of hundreds of production Fords, Lincolns, and Mercurys, Edsel, in his typical self-effacing manner, was quick to point out that he was not a designer. "I can judge design," he once said in a rare interview, "for on that I have had long experience. But I can only judge." George Pierrot, a long-standing member of the Detroit Institute of Arts and Edsel's close friend, agrees with Edsel's assessment of himself. "Edsel possessed unusual sensitivity and a rare feeling for form, composition, and color," Pierrot explained. "His taste and critical judgment were extraordinary."

Edsel's ability to discern an ordinary line from a classic one, as Ray Dietrich described it, made Edsel one of the most talented yet uncelebrated design directors of his time. It would be easy to imagine that if Edsel had not been president of his own automobile company, he could easily have been the design director of Chrysler, Hudson, or Packard. "Edsel had a sense of what looked right," says Gregorie. From his years of drawing automobiles and

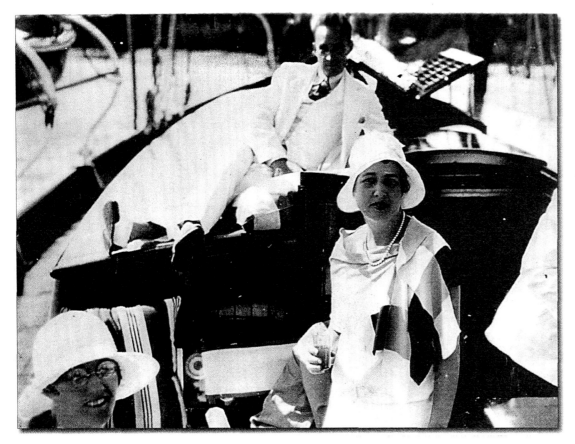

Figure 3-23. Edsel Ford and his wife, Eleanor, aboard their yacht, Acadia, *circa 1927. Edsel loved boats, and he encouraged E.T. Gregorie to adapt their simple, straightforward lines to the design of Ford products. (Photo from the Collections of Henry Ford Museum & Greenfield Village)*

to implementing Alfred Sloan's marketing strategy. His goal was to offer a vehicle for "every purse and purpose" and to stimulate the demand through yearly design changes. These changes were not always extensive. Most were merely facelifts, modifying the grille or the rear end of the car only enough to make the models from the previous year obsolete.

Although Edsel persuaded his father to make a few cosmetic changes to the Model T over the years, these changes were neither frequent enough nor significant enough to maintain product excitement in the mind of the consumer. By the time the last Model T was produced in 1927, an entirely new car was needed. "Edsel's influence came in when all the companies started coming out with yearly changes to their models," Joe Galamb explained. "He was more aggressive, and he knew what he had to do to keep in line with these competitors and not fall back like [the company] did with the Model T. [Edsel] liked the idea of how General Motors handled their bodies and chassis—the little changes at the front end and the changes that did not affect the general design of the doors and roofs. He wanted to get that idea carried through the Ford Motor Company. He had very good taste and a sense of market demand and preference." Therefore, starting with the Model A, Edsel's interest in design was accelerating, and he began emulating General Motors.

studying their design, Edsel knew intuitively what was a good design and what was a poor design, and that is what made him such a capable "styling executive."

The design of the Model A was, in many respects, a knee-jerk reaction to the competitive environment that confronted Ford Motor Company in the late 1920s. By 1927, the year in which the Model A was designed, General Motors was well on its way

Figure 3-24. Edsel Ford drives his speedboat, Goldfish, *on the Detroit River, circa 1916. (Photo from the Collections of Henry Ford Museum & Greenfield Village)*

When the design of the later Model A was complete, Edsel immediately ordered another complete redesign. Again, this was sparked by Chevrolet. To help thwart the success of the Ford Model A, Chevrolet introduced a six-cylinder car in 1929. Not to be outdone again, Henry Ford developed the first mass-produced V-8 engine and put it in an entirely new Ford for 1932. (Figures 3-25 through 3-27) Again, Edsel took over the design of the new car, but this time he turned to Briggs Manufacturing to help him.

Briggs had been building bodies for Ford since the Model T days but was never involved with design until Edsel insisted. "You do a substantial part of our body business—and that's about one third of the entire car—yet I pay for all the design work and advertising," Edsel complained to Walter Briggs. "In order for us to remain competitive, we must continue to advertise and develop new designs, but we cannot do both ourselves. You need to provide design expertise."

Consequently, as soon as the Model A went into production, Edsel and Galamb began working on a facelift for the car, which was implemented in 1930. With its rounded radiator shell, sharply curved fenders, and vertical windshield, the 1928–1929 Model A was cute, but the new version had a more refined and substantial appearance. The 1930–1931 Model A had a taller radiator shell, a longer-looking hood, and more gracefully curved fenders.

Concerned that Edsel would pull away from his company if he did not follow his edict, Walter Briggs convinced Ralph Roberts in 1927 to merge LeBaron with Briggs Manufacturing. It was Roberts' expertise and Edsel's discerning eye that led to the design of the handsome 1932 Ford line. In 1932, Roberts hired John Tjaarda, a capable designer with roots in custom-body design, and the two of them hired some of the most talented automobile designers at the time: Phil Wright, Jim Wilson, Alex Tremulis, Holden Koto, Bill Flajole, Walt Wengren, Clarence Karstadt, Al Prance, and a young Rhys Miller. (Figure 3-28) Soon, the Briggs design studio was second in size only to the Art & Colour Section of General Motors.

With this powerhouse of talent at his disposal, Edsel began mimicking Harley Earl's yearly model change philosophy. Holden Koto and Phil Wright, under the direction of John Tjaarda, helped Edsel design the all-new 1935 Ford, the facelifted 1936 Ford, and most of the 1937 Ford. "These cars with their rounded window openings and teardrop-shaped fenders are indicative of Tjaarda's work," explains Rhys Miller, a design apprentice who started working at Briggs in the spring of 1935.

However, the stunning 1933 and 1934 Fords were based on a design developed by Ford's own E.T. Gregorie.

Figure 3-25. This early version of the 1932 Ford design was proposed by Briggs Manufacturing. Edsel thought the snow-shovel front end was too radical. (Photo from the Collections of Henry Ford Museum & Greenfield Village)

Figure 3-26. In this later version of the 1932 Ford design developed by Briggs, the body and radiator grille design are close to how they eventually went into production. Edsel was instrumental in directing designers from Briggs toward this cleaner, more conservative design. The date on the photograph is misleading; it is the date the photo was taken by Ford photographers, not the date the model was built. (Photo from the Collections of Henry Ford Museum & Greenfield Village)

Figure 3-27. This is the production version of the handsome, conservatively designed 1932 Ford. (Photo from the Collections of Henry Ford Museum & Greenfield Village)

Figure 3-28. This rare photograph shows designers at Briggs Manufacturing, circa 1933. The man in the dark vest is the legendary designer Ralph Roberts, design director for Briggs and former partner in LeBaron, Carrossiers, which designed numerous custombodied Lincolns for Edsel Ford. To his left is Al Prance, who took over the Briggs design studio after World War II. The designer wearing glasses in the foreground is Jim Wilson. The other two designers are unidentified. Roberts and his colleagues are examining design renderings of proposed 1935 Fords. These men, along with John Tjaarda, Phil Wright, and Holden Koto, designed the 1932, 1935, and 1936 Fords, and part of the 1937 Ford. (Photo from the Collections of Henry Ford Museum & Greenfield Village)

CHAPTER FOUR

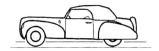

THE FORD MODEL Y
Gregorie's First Ford Design

In the summer of 1931, a few short months after Gregorie had started with Lincoln, Edsel Ford received disturbing news from Lord Percival Perry, Ford's European manager, who said that the Model A was not selling well in the British market. In those depressed times, the Model A was too expensive to buy, too expensive to operate, and too large for the narrow streets of Europe. Usually, Ford products were competitively priced in all markets in which they were sold, but that was not true in England. On January 1, 1921, the English Finance Act went into law, placing a tax of one pound sterling per horsepower on all passenger cars. Interestingly, the horsepower was determined by an odd formula developed by the RAC (Royal Automobile Club), which placed substantial significance on the number and diameter of cylinders in an engine. Therefore, a motorcar with fewer, smaller cylinders was taxed less than one with more, larger cylinders. (The stroke of the engine was not taken into consideration.) Because the Model A had four large-diameter cylinders, it was assessed a heavy tax—one large enough to dramatically affect sales. What Perry wanted was a new car, designed specifically for the British market.

Shortly thereafter, two of Perry's men visited Dearborn, and they met with Edsel Ford and Larry Sheldrick, the chassis engineer at Ford, to outline their requirements. The new car, they said, "must have a small bore engine of about eight [RAC] horsepower, a ninety-inch wheelbase, a narrow tread, and limited weight." In addition, the design must be done as quickly as possible.

Sheldrick went to work immediately on a new engine and chassis for the new British car, which company officials dubbed the Model Y. Curious to see what Gregorie could do on a small car, Edsel asked him to design the body. This would be Gregorie's first chance to design a complete body for a Ford car and prove his design prowess to Edsel.

Because of the tight schedule and undoubtedly because of Henry Ford's watchful eye, Sheldrick essentially took the Model A engine and chassis and "shrunk" them to fit the dimensions outlined by Perry's engineers. The engine was literally half the size of the Model A engine in all dimensions, and the tiny bore of its four cylinders ensured that it would calculate out to eight RAC horsepower. The chassis was all Ford: buggy-type transverse springs front and back, torque tube drive, and solid front axle. However, the body design was unique and especially handsome, particularly on such a small chassis.

According to automobile designers, it is extremely difficult to make a small car attractive. "It all has to do with proportion," explains Gregorie. A small car must accommodate an average-size person as a large car does, but when that reality is incorporated into a small package, the design always appears top-heavy and stubby. Gregorie faced these predicaments while designing the Model Y. However, according to Gregorie, "Edsel asked me to see what I could do with a car of this size and asked me to go over and see Sheldrick to get the specifics."

When Gregorie walked in, Sheldrick was not there, but his assistant, Eugene Farkus, was present. Farkus had been with Ford since 1908 and had helped design mechanical parts for the Fordson tractor and Model A. He was assisting Sheldrick in the chassis design for the new Model Y when Gregorie entered. Farkus still remembered Gregorie's visit years later. "I got acquainted with a young fellow who got started [with Ford] a month or two before," Farkus recalled. "His name was Gregorie...and he showed me some advanced sketches of different body designs he had developed." Farkus looked over Gregorie's sketches and exclaimed, "You're just the man we want. We want you to help us out on this little English job. Design us a nice, up-to-date body for it."

"That's what Mr. Ford asked me to do," replied Gregorie.

Gregorie obtained a blueprint of the chassis from Farkus and began developing a design in scale. "I was told to design a four-door and a two-door," explains Gregorie. (Figure 4-1) Starting from the front of the car, the first thing Gregorie did was slant the grille shell rearward. Then he carried that theme to the windshield, slanting it as well. Next, he extended the hood line so that it came right to the windshield. "I did that," he says, "to give the car a longer effect. It eliminated the joint line between the hood and the cowl, and that gave the car the appearance of more length, which was important on a car of this size. That feature soon became the standard of the industry."

In addition to slanting the grille, Gregorie gave it a heart-shaped design, with fine, slightly concave vertical bars. He eliminated the headlight bar—a styling cue that other manufacturers adopted the following year—and designed the front doors to open rearward. "It was my idea to have the center-hinged doors," explains Gregorie. "That gave us two things. It's a lot easier to get in and out since it gives you more leg room, and we were able to get a nice slant to the windshield which added to the gracefulness of the design. Both of these things are important in a car of this size." Gregorie added a valence to the front fenders—another innovative styling cue—and gave them a nice, pleasing slope. To match the flow of the front fenders, Gregorie flared the rear fenders to "support" the rear of the body. "We designers call that a cheat line," explains Gregorie. "The extended line of the rear fenders gives the illusion of more length."

After Gregorie completed the design in scale, he recalls that "it was all drawn up full size on a blackboard, and then an engineering draft was made. These drawings were then sent over to England. A full-size wooden model of this car was eventually made, but it must have been made by the Briggs people in their plant in England—they had a stamping plant next to Ford's Dagenham plant. I know we didn't make the wooden model in our shop." (Figure 4-2)

What Gregorie did not know at the time is that Ralph Roberts, head of the Briggs design studio in Detroit, heard about Edsel's desire for a new English model and pursued some ideas for the little car on his own, hoping he could sell one of them to Edsel. When Roberts had a design ready, he took it to Dearborn to show Edsel. After reviewing what Roberts had done, Edsel tried to let Roberts down easily. "I wouldn't care for it," Edsel explained, "but if that's what Perry wants, okay."

Figure 4-1. Gregorie's first design triumph was the Ford Model Y. It was a tremendous success in England, the market for which it was designed, and was produced between 1932 and 1937. (Photo from the Collections of Henry Ford Museum & Greenfield Village)

Figure 4-2. This photograph of the full-size wooden model of Gregorie's Model Y design was taken inside the Ford engineering laboratory, but the model was built by Briggs personnel in the Briggs design studio. (Photo from the Collections of Henry Ford Museum & Greenfield Village)

Taking Edsel's cue, Roberts headed for England, hoping Perry would buy this design and give Briggs more business. His long trip was for naught, because Perry did not care for his design any more than Edsel had. "That's not what we want," Perry explained. "But if that's what Edsel wants, I guess it will be all right." With that, Roberts knew he had lost the secretive contest. He left England empty-handed, and Gregorie's design was the one that went into production.

Roberts's sketches were lost long ago, but they would have had to have been stunning to beat Gregorie's design. Gregorie admits that he used the 1932 Ford as a baseline for his own concept. That car was handsome, but Gregorie gave it some graceful lines and additional style. In other words, it was a smoothed-up 1932 Ford. "It leapfrogged over the '32 model," says Gregorie.

Figure 4-3. This Gregorie-designed and German-made 1938 Eifel was designed a full two years before Gregorie's handsome 1939 American Ford models, but Gregorie gave his new Model Y lines that he would later incorporate into his American Ford designs. (Photo from the Collections of Henry Ford Museum & Greenfield Village)

When production of the Model Y began, Lord Perry sent a few cars to Dearborn at Edsel's request. There were several Tudors, a couple of Fordors, and a cute little cream-colored touring car that Perry had built especially for Edsel. "I liked these cars, they were smart little cars," explains Gregorie. "I got hold of two of them, one was a Tudor and the other was that little touring car. Edsel gave me both of them."

Years later, Gregorie redesigned the Model Y, first along the lines of the 1937 Ford and then again along the lines of his 1939 Deluxe Ford and new Mercury. (Figure 4-3) These cars were produced in Germany as the Taunus and in England as the Prefect. Val Tallberg, Ford's plant manager of the Cologne Works in Germany in the late 1930s, "squeezed" a couple of the new Ford

Taunus vehicles out of Germany through Sweden and brought them to Dearborn immediately before the onslaught of World War II. "I grabbed one of them as soon as I could," says Gregorie. "I kept all three of them—the Model Y Tudor, the special touring car, and the Taunus—at my house and owned them until 1945. I drove them all during the war. Gas was scarce during the war years, and with their tiny engines, they were wonderful on fuel. I eventually sold the touring car to a Ford dealer in Philadelphia, who wanted it for his daughter."

When asked if the Model Y was a test case given by Edsel to see if he had the capability of designing a complete car, Gregorie replied, "I don't know if Edsel was testing me or not, but it was the first full car that I designed for Ford." In any case, whatever Edsel may have thought, Gregorie's little Model Y was a big hit with Europeans, and Lord Perry was elated with its success. Over a nineteen-year period, more than 600,000 copies of Gregorie's design were produced. After World War II, the Taunus version continued in production until 1951.

CHAPTER FIVE

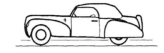

THE 1933 AND 1934 FORDS
"Like a Man's Face"

Soon after Gregorie had completed the design of the Model Y, Edsel began talking to his father about developing a new Ford line for 1933. The 1932 Ford had been a crash program, from the chassis and new V-8 engine developed by Henry, to the body design developed by Edsel and Galamb. However, as far as Edsel was concerned, it was only an interim step toward building more graceful motorcars. The 1932 Ford remained too stubby for his liking. For the 1933 Ford, he wanted a more graceful design on a longer chassis.

Henry Ford was reluctant. He was not used to making major changes so often. He thought the Model A could have been produced for years without any major changes, in the same way that his Model T had been. He made a change in 1932 only because of pressure from the marketplace and his desire to "put the competition in their place" by developing the first low-cost V-8. Fortunately for Henry, his son also developed a handsome body design to complement the new engine and chassis. Now that Henry had allowed two major changes in four years, Edsel was confident that he could convince his father to make changes more frequently in the future.

For the 1933 line, all Edsel wanted was a longer wheelbase. After a little consoling, his father agreed to add six more inches to the wheelbase of the 1932 Ford. Although not a significant amount for the engineers to change, the added length helped Edsel tremendously in developing automobiles with a more graceful appearance. Edsel's challenge now was to develop a suitable design, and to do it quickly.

Edsel pondered the possibilities, wondering whether the lines of Gregorie's handsome Model Y could be adapted to the longer chassis. Edsel and Gregorie had not yet developed a formal working relationship; therefore, instead of discussing the idea with Gregorie, Edsel relied on his usual process of working out designs with his body engineers. He asked Clare Kramer, a body draftsman who worked for Joe Galamb, to take the lines of Gregorie's Model Y and "proportion them up" to fit on a chassis having a wheelbase that was a full twenty-two inches longer than that of the Model Y. A week or so later, Kramer cornered Edsel and showed him the profile that he had developed. It was a stunning transformation. The larger proportions of the 1933 American Ford transformed the proportions of Gregorie's handsome Model Y into a strikingly beautiful automobile. "Roll it up for me," instructed Edsel, pointing to the sketch. "I want to take it with me."

With the drawing in hand, Edsel went immediately to Gregorie's drafting board and showed him the design. Edsel unrolled the drawing on Gregorie's drafting board and asked, "How do you like it?"

"Fine," Gregorie replied.

"Do you recognize it?" Edsel asked.

"Yes, of course, Mr. Ford," Gregorie answered.

The entire thing completely surprised Gregorie, but he was proud of the outcome. "Engineering handled the whole process, at the order of Edsel Ford," explains Gregorie. "They stepped up my Model Y design a certain percentage to produce the smart line for the '33 and '34 Fords."

What a smart line it was. (Figures 5-1 and 5-2) It became one of the most beautiful lines of cars in automotive history. Although Gregorie was not involved in the design process of this line of Fords, it was nevertheless *his* lines and *his* design copied by the body engineers.

Figure 5-1. The stunning 1933 Ford was merely a larger version of Gregorie's popular Model Y and became one of the most beautiful cars ever designed at Ford. (Photo from the Collections of Henry Ford Museum & Greenfield Village)

Figure 5-2. Edsel Ford (left) proudly displays a 1933 Ford to his father (right) in December 1933. (Photo from the Collections of Henry Ford Museum & Greenfield Village)

Looking back over six decades, Gregorie becomes teary-eyed when looking at photographs of the 1933–1934 Fords. "Gee, those are good-looking cars," he says. "The general proportions of those cars are excellent. They are nicely balanced between the front end and the rear end. They have nice detail; everything had a purpose. But most important, the front wheels and the rear wheels are as far fore and aft as possible, and the front axle was ahead of the grille, and that is the basis for a good design. Unfortunately, the frame is high, and by the time you put the platform on top of the frame, and the seats on top of the platform, you are up to your bosoms in car! Interestingly, that's one of the redeeming things about the cars they're building today: they have a low profile. But unfortunately, a low profile has its disadvantages as far as headroom and entry and exit are concerned. With the '33 and '34 Fords, you can slide in without any difficulty. In today's cars, you have to double up, put one foot in at a time, and slide sideways in order to get in. You have to be a contortionist!

"The old cars also become attractive to people today because they literally have an expression," continues Gregorie. "When you had separation of fenders and headlamps and bumpers, each one was a piece of artwork in itself. And when you put them all together, the assembled front end gave the car unique composition and expression, like a man's face—the bumper being the mouth, the grille the nose, and the headlamps the eyes. Some look spunky, some look sassy, some look terrible, but they all had their own identity."

The Model Y represented a peculiar chapter in Ford development. "It was a strange way it came about," Gregorie recalls. "People find it hard to believe that the full-sized '33 Ford was modeled on that small version. As it turned out, it saved a hell of a lot of development time—just step the dimensions up, and it turned out smooth and nice. The '33 and '34 were two of the nicest-looking Fords I had anything to do with. They were real graceful and nice cars, especially the one I had built for myself."

Starting with a standard Ford roadster, Gregorie had some expert tradesmen at the aircraft factory make a number of alterations to the black roadster with red leather interior. They chopped the top, slanted the windshield, turned some beautiful aluminum disc wheel covers, and fabricated some handsome fender skirts. "It was beautiful," says Gregorie. Another unique feature of his little roadster was its honeycomb radiator grille. "I don't know how the hell they made it, but it was a die casting. It was sent in by some supplier as a sample to use on the '34 Ford. We'd normally just throw that kind of stuff away, but I had a storeroom in the back of my office where I kept a lot of those experimental parts. Anyway, I pulled that radiator grille out and had the aircraft boys stick it on my roadster. Everybody scratched their heads, wondering where in the hell it came from. They probably thought I had it made especially for myself at the Ford shop."

Throughout his career at Ford, Gregorie wished that Henry Ford would make the chassis longer and lower. That is the only way to design beautiful automobiles, and the transformation of the short, squatty Model Y into a longer, lower-looking automobile proved that point. Unfortunately, the Ford chassis became only two inches longer the entire time Gregorie designed cars at Ford. As it turned out, that additional length was added in the wrong place.

CHAPTER SIX

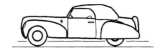

THE CONTINENTAL CAR I
"Long, Low, and Rakish"

Between 1912 and 1939, Edsel Ford made several trips to Europe. His first trip was taken with his father and mother to commemorate Edsel leaving high school and joining the company on a full-time basis. Most of these trips to the "Continent," as Edsel called Europe, were for pleasure. However, Edsel always managed to work in some business by visiting the various Ford plants in England, France, and Germany. What he enjoyed most during these trips was not visiting the Ford plants or the antiquities of Europe, but seeing the continental cars. To Edsel, European motorcars were more elegant and graceful than American cars, and it became his dream to someday add continental class to the cars he made in Dearborn.

After returning from a trip to the Continent in 1932, Edsel rushed to see Gregorie and began discussing with him the possibility of adding a continental-type car to the Ford line. They talked about various kinds of cars found on the Continent, but particularly about racy sports cars such as Bugattis, S.S. Jaguars, Lancias, and Alfa Romeos. In the end, they decided to pursue a design that was, as Edsel put it, "long, low, and rakish." However, those were the only attributes mentioned at the time. The two men did not discuss *how* long or *how* low the car should be, its size, or even its appearance. The only clear-cut direction that came from their meeting was that the quickest and cheapest way to place such a new model into production would be to simply design a sporty body and add it to a standard Ford chassis. That is the approach Gregorie pursued.

To assist Gregorie in his efforts, Edsel gave him complete use of the service department at the Ford aircraft factory, located down the street from the Ford Engine & Electrical Engineering Building (known to everyone as the engineering laboratory, or simply the engineering lab). With the onslaught of the Depression, orders for Ford Tri-Motor airplanes had dropped to a trickle, but the company kept a cadre of expert tool and die makers, metal fabricators, upholsters, and other craftsmen to service and repair the all-metal planes. If necessary, these craftsmen could fabricate from scratch almost any part a plane needed, and Gregorie utilized them to assist his design efforts. "The aircraft plant had a large surface plate," recalls Gregorie, "big enough to place a couple of cars on to check dimensions and so on. That is where we built these special cars."

Inspired by his love of boats, lines of the classics, and popular styles of the day, Gregorie completed his first attempt at designing a continental-type car for Edsel in the fall of 1932. (Figure 6-1) Made completely of aluminum by the aircraft men, the boattail roadster was built on a standard 1932 Ford chassis. The sleek body had cycle fenders, bullet-shaped headlamps, sporty disk wheels, and a V-shaped, raked windshield. Its front end was similar to that of a standard 1932 Ford, but Gregorie modified the grille into a slight V shape and carried that trait back to the two-piece windshield. As previously agreed, the chassis and drive train were standard Ford components.

Edsel was quite impressed with Gregorie's design, and he drove the little roadster around his Gaukler Pointe estate for several months afterward, getting a feel for how it drove and soaking up its design. Unfortunately, other pressing matters, such as running the company and dealing with the effects of the Depression, sidetracked Edsel from further pursuing the continental project until the summer of 1934.

In February 1933, the Depression had reached its nadir. Banks all over the country were closing, including a large group of banks in Detroit headed by the Guardian Group, of which Edsel Ford was a major stockholder. To help quell the situation, Michigan Governor William Comstock issued a banking moratorium on February 14, instructing all banks in the state to remain closed for a week. Before the week had ended, the governor extended the moratorium another six days, and then another four days. On March 5, President Franklin Delano Roosevelt, as his first official act, was forced to close every bank in the nation.

With the restriction in the money supply and with automobile sales being almost nil, Ford Motor Company all but closed, laying off thousands of people. Included in these layoffs were Gregorie and the others working on Lincolns. From February 1933 through April 1934, no design activities occurred at Ford Motor Company. For all intents and purposes, the entire company had shut down. "That was the first time I left Ford," recalls Gregorie. "With time on my hands, I decided to take a trip to England. For a hundred and eighty-nine dollars, I bought round-trip passage on the *American Merchant*, a freighter that also carried about thirty passengers, and spent the summer in England. Percival Perry provided me with a Model Y, and I was able to see a lot of England's countryside that way. While I was there, I made acquaintances with people at Rootes Body Company and also at Hooper & Company—two custom-body firms in the London area—and did some design work for them after I returned home to America that fall, getting passage on the *American Merchant's* sister ship, the *American Banker*. I didn't do any work for them while I was on the Continent, since my visa would not allow it."

From the fall of 1933 through the spring of 1934, Gregorie spent time with his family in North Carolina. Finally, in April, Gregorie was called back to work. Instead of returning to his desk in the engineering laboratory, Gregorie was instructed by Edsel to set up a drafting board in the company's airport terminal.

When Edsel finally convinced his father to go into the aircraft manufacturing business in 1924, they entered the business with a fury. They cleared a 700-acre section of land south of the engineering laboratory and built an airport with concrete runways, an expansive manufacturing plant, hangars, and a passenger terminal. The terminal was a two-story structure, with offices on the second floor, and the passenger lounge and ticket office on the first floor. (Figure 6-2) To accommodate passenger traffic, the 100-room Dearborn Inn was built across the street from the passenger terminal. "I ate lunch at the Dearborn Inn quite often during that time," Gregorie recalls. "In fact, I lived there for a short time after I returned from England. The night manager there, who was a friend of mine, gave me a room for twenty-eight dollars a week, and that included a heated stall for

Figure 6-1. This boattail speedster is the first continental car Gregorie designed and had built for Edsel Ford. Utilizing a standard 1932 Ford chassis and power train, the special body and fenders were handmade at the Ford aircraft factory, shown in the background. (Photo from the Collections of Henry Ford Museum & Greenfield Village)

Figure 6-2. The Ford airport terminal was all but abandoned with the cessation of Tri-Motor airplane production in 1933. There, in this air-conditioned facility, Edsel set up Gregorie in 1934 so that Gregorie could design cars in comfort and out of view of Henry Ford. This was the start of the design department at Ford Motor Company. (Photo from the Collections of Henry Ford Museum & Greenfield Village)

my Stutz in the basement garage! Whenever I brought it in, they'd wash it for me. Well, those were hard times, and the manager was glad to have anybody staying there."

Until the time Gregorie was laid off, he had been primarily working on Lincoln designs for Henry Crecelius, chief body engineer at Lincoln, and only dabbling in Ford designs now and then for Edsel. However, now that Gregorie was ensconced in the airport terminal, he knew things were about to change. Edsel told him, "You'll be working directly for me."

Gregorie liked that arrangement because it was informal. "I set up a drafting table on the main floor," he recalls, "and Edsel would stop in to discuss designs that I was working on, without having to go over to the engineering lab. He didn't have me work on anything specific at this time. He merely had me develop ideas." The work setting was informal indeed, especially in summer. When it was unbearably hot and humid outdoors, Gregorie was comfortable in his air-conditioned work area. Often, he removed his shoes and worked barefoot, standing at his drafting board on the cool terrazzo floor. "I guess that's one way to be comfortable," Edsel would say.

As rudimentary as it was, this was the beginning of the in-house design department of Ford Motor Company.

CHAPTER SEVEN

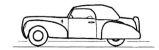

THE CONTINENTAL CAR II
Lower and Longer

When Gregorie became settled in his new environment, he and Edsel again began talking about a Ford sports car. Edsel liked Gregorie's first boattail design, but he wanted Gregorie to design and build another one that was much longer and lower.

Gregorie began immediately. He sketched a few ideas and finalized his favorite into a one-tenth scale model, testing it in the wind tunnel at the aircraft factory to "see how the air flowed around it." In this model, Gregorie envisioned a car that was substantially longer, lower, and much more streamlined than his first attempt. The overall design was unique, but the front grille work was particularly distinctive. Gregorie divided the grille into two pieces: (1) a vertical, V-shaped upper portion, and (2) a horizontal, oval-shaped lower portion. The lower grille also housed the headlights, and the entire assembly was molded into the body, a unique feature at the time.

Gregorie showed Edsel the model and gave him a photograph of it to ponder. (Figure 7-1) Overall, Edsel liked the new design but wondered "if the two grilles shouldn't join?"

Taking this subtle cue, Gregorie revamped the front end into a one-piece, V-shaped grille. From there, the hood was stretched all the way back to a truncated, two-piece windshield. He separated the headlights from the grille but left them molded into the body. Then he covered the specially made disk wheels (which had handmade hubcaps with the Ford "V-8" symbol embossed on them) with cycle fenders made from fenders for a Tri-Motor airplane. The front fenders steered with the front wheels. "We had an expert metal beater," explains Gregorie. "He'd beat those fenders out on a big leather bag filled with shot. He'd take that leather bag, stand over it, and pound and pound the metal on it. He'd learned that skill in the old country—making pots and pans by hand. Man, he'd beat, beat, beat, beat, hour after hour, and he'd shape that metal around into the most beautiful shapes by hand." (Figure 7-2)

To achieve the long, low effect that Edsel wanted, Gregorie was forced to fabricate an entirely new frame for the sports car. He started with a standard Ford frame, but by the time he and "the boys" at the aircraft plant finished chopping, stretching, and welding it, they probably would have been better to build a special frame from scratch. It took them almost as long to fabricate the frame as it did the body.

Edsel drove this new continental car around his Gaukler Pointe estate for many years, but he and Gregorie met to discuss the salient features of this new design shortly after Edsel took possession of the car in the summer of 1934.

By this time, Edsel and Gregorie had spent a lot of money designing and building these experimental continental cars. Estimates place their cost in excess of $100,000 each to produce, but the two men had gained significant experience in the process. In fact, the parameters of a new Ford continental car had finally, if not accidentally, been determined. Gregorie's latest body design was too streamlined for Edsel's taste, but its long, low stance was exactly what Edsel was seeking for a continental car. Perhaps by combining his latest chassis with a more conservative body style similar to his first design, Gregorie could develop an acceptable continental car. That would be the tack he would take when he returned to the continental car project. However, before that could happen, Gregorie would have to complete a new assignment from Edsel.

Figure 7-1. Before starting the second continental car for Edsel, Gregorie made a number of sketches, finally settling on this design, which he made into a model. Edsel examined the model and, on this photograph of it, wrote some subtle comments for Gregorie to consider. (Photo from the Collections of Henry Ford Museum & Greenfield Village)

Figure 7-2. This is the final version of the second continental car Gregorie designed and had built for Edsel Ford. To achieve this long, low effect, Gregorie radically modified a standard Ford frame. The cycle-type fenders were made of Ford Tri-Motor airplane fenders—the front ones steered with the wheels. Picking up this car from Gregorie in the summer of 1934, Edsel drove it around his Grosse Pointe estate until his death in 1943. (Photo from the Collections of Henry Ford Museum & Greenfield Village)

CHAPTER EIGHT

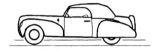

THE 1936 LINCOLN-ZEPHYR

"The First Successfully Streamlined Car in America"

Throughout the 1930s, Edsel Ford took steps to position Ford Motor Company to compete one-on-one with General Motors. At the start of the 1930s, General Motors had five model lines, and Ford had two. By the end of the 1930s, General Motors still had five lines, but Edsel had increased Ford's offerings to five lines also: Ford, Ford Deluxe, Mercury, Lincoln-Zephyr, and Lincoln. The Ford Deluxe was simply a fancier Ford, whereas the Mercury, introduced in 1939, was an entirely new automobile. Although the Ford and the Mercury appeared to belong to the same family, no body panel from one car would fit onto the other. The Lincoln, however, was in a class by itself and resembled nothing but a Lincoln.

The Lincoln-Zephyr, also a unique Ford offering, was the first new model line in the Ford family since the company had acquired Lincoln almost fifteen years earlier. However, it was not a Ford-developed design; rather, it was the brainchild of John Tjaarda, a talented designer hired by Ralph Roberts at Briggs Manufacturing. The Zephyr began as an offshoot of the streamlined designs that Tjaarda had been developing during the previous ten years, which he called Sterkenbergs—after his own original family name. (Figure 8-1) These designs had many innovative features, including unit-body construction, severely sloping front end, fastback rear end, and rear-mounted engine.

Walter Briggs, well aware of Edsel's desire to expand the Ford product line, sold Edsel on the idea of offering Tjaarda's Sterkenberg design as a competitor to Chrysler's Airflow, which was introduced in 1934. Edsel was intrigued with Tjaarda's design when he first saw it, but he also realized that, to successfully market the streamlined car, the engine would have to be moved to the front and its blunt front end redesigned. (Gregorie called it a "sheep's nose.")

Edsel asked Briggs to have Tjaarda develop some ideas that incorporated those changes, but Tjaarda dragged his feet. He wanted his Sterkenberg design to go into production without alteration, and every new proposal that he showed Edsel incorporated essentially the same design. Consequently, Edsel vetoed every new

Figure 8-1. Tjaarda's Sterkenberg design was displayed at the Ford Exposition of Progress in Detroit in 1933. Edsel bought this design from Briggs Manufacturing and, after Gregorie revamped its "sheep's-nose" front end, transformed it into the highly successful 1936 Lincoln-Zephyr. The sign above the car reads: "This car is our design contribution and conception of a stream-lined automobile and has no reference to current or future 'Ford' production—Briggs Mfg. Co." (Photo from the Collections of Henry Ford Museum & Greenfield Village)

idea that Tjaarda presented to him. As time and ideas passed, Edsel became discouraged and doubted whether the design could be saved.

While Tjaarda was making excuses, Edsel was ruminating the problem and eventually had an idea that, on the surface, seemed completely nonsensical. Amazingly, Edsel visualized that a front end with sharp, vertical lines could be mated handsomely to Tjaarda's bulbous design. He never mentioned his idea to either Tjaarda or Briggs, partly because he knew they were trying to retain the original design of the car, and partly because he did not want to be challenged. Instead, he approached Gregorie with his idea, knowing that Gregorie would be more tolerant of his suggestion.

However, Gregorie too was flabbergasted. He could not believe what Edsel was telling him and thought that perhaps his patron had suddenly lost his sense of styling! To him, Edsel's approach seemed contradictory to the purpose of aerodynamics, and Gregorie tried to persuade Edsel from pursuing such a nonsensical approach. "But Edsel was very obstinate," says Gregorie. Edsel truly felt that, if done properly, a pointy shape could be married beautifully to Tjaarda's design, and he told Gregorie to go to the Briggs studio and "help Tjaarda out."

Indeed, that was an awkward position for both Tjaarda and Gregorie. However, when Gregorie arrived at the Briggs studio, he pulled Tjaarda aside and had a heart-to-heart talk with him about the situation. "To be honest with you, John," he said, "I really don't care much for this entire project, but Mr. Ford wants us to come up with a front end that will be acceptable to get this car into production. Now, let's cooperate on this, and perhaps we can come up with something that is acceptable to Mr. Ford."

Realizing that there was no benefit in stalling any longer, Tjaarda finally acquiesced, and he directed Gregorie to an open drafting table. "So I sat there, and on the back of a blueprint, I sketched up a front end in perspective," recalls Gregorie. "Edsel loved pointy shapes, and he wanted a pointy grille on this car. So I sketched up this pointed front end with a kind of inverted boat hood." Starting from the top of the cowl of Tjaarda's design, Gregorie sketched in an almost perfectly horizontal line forward, which became the length of the hood. Then he drew in another line downward at a severe, almost vertical angle, which became the leading edge of the grille. When he was through, he had shaped in a pointy grille as Edsel had wanted. Weeks had passed since Edsel first asked Briggs to develop a new front end design, to no avail. However, Gregorie needed only thirty minutes to create the design that saved Tjaarda's car.

As Gregorie was sketching the new front end, he thought it would be too angular and too pointy for a streamlined car, but when the two seemingly discordant forms were mated, they "clicked, clicked, clicked." No one was more surprised than Gregorie. According to Larry Sheldrick, who had witnessed the entire Zephyr project, Gregorie "made his reputation with Edsel by refining and completing the styling begun by Tjaarda," and that ultimately paved the way for Gregorie to becoming the first design chief at Ford Motor Company.

Edsel was pleased with Gregorie's enhancements, but he was not surprised that the two supposedly discordant shapes fitted together so well. After all, that is what he had visualized. Thus, the front end of Tjaarda's design was "changed accordingly for a quick okay," and the car went into production as the Lincoln-Zephyr in the fall of 1935, as part of the Lincoln line. (Figure 8-2) Selling at $1,275, the car invaded the upper-medium-priced field, which is precisely what Edsel wanted. It essentially gave Ford dealers a third car to sell, in addition to the standard Ford and the upscale Lincoln. Not evident at the time, the Zephyr would also provide the foundation for Gregorie's most famous creation.

Although Gregorie saved Tjaarda's design, he was never fond of the Zephyr. "I don't think it was one of our best design efforts," he confesses, "but it sold the car to Edsel Ford, and that is what I was being paid to do. Actually, I think the car probably would have been more appropriate with the rear engine and sheep's-nose front end, as Tjaarda had originally conceived it." (Figure 8-3)

Figure 8-2. Shown here is the production 1936 Lincoln-Zephyr. Gregorie's pointy front-end treatment on this car saved Tjaarda's concept from oblivion and gave Edsel another model to sell. (Photo from the Collections of Henry Ford Museum & Greenfield Village)

The 1936 Lincoln-Zephyr

Figure 8-3. Shown here are two Lincoln-Zephyr design proposals done at Briggs Manufacturing subsequent to Gregorie's more conservative approach. Gregorie prefers the one on the bottom. "It looks pretty sporty," says Gregorie. "It looks like it's ready to get up and go." Neither of these front ends made it to production. (Photos from the Collections of Henry Ford Museum & Greenfield Village)

Gregorie is equally critical of the engineering of the car. "The Zephyr wasn't a good car on the road," he says. "It had poor weight distribution. It was too nose heavy, and it was too light in the rear end. I told engineering one day, 'If you would've put a nice, big straight-eight engine in this car—you know, a good lugging engine like what the Buick has—we'd really have something!' The Zephyr was beautiful to look at and was beautifully finished, but it wasn't the best piece of engineering in the world."

The Lincoln-Zephyr continued in production until February 1942 when all Ford civilian production stopped as the United States entered World War II, but it was never reinstated when Ford resumed automobile production in 1945. Between its introduction in 1936 and its demise six years later, the Zephyr went through several styling changes—the most dramatic in 1938—but the company decided not to offer it after the war.

In 1951, five years after Gregorie left the company, the Museum of Modern Art in New York City held "an exhibition concerned with the aesthetics of motorcar design." In the exhibit, the museum critiqued a number of automobile designs. Among them were the Cord, the Bentley, and the Cadillac 60 Special. The curators said the Cord "suggests the driving power of a fast fighter 'plane"; the Cadillac "a five-passenger sedan of restrained design"; and the Bentley "an elaborate box with curving planes and sharp edges." However, they called the 1936 Lincoln-Zephyr "the first successfully streamlined car in America." The Lincoln-Zephyr, they said, "has an impeccable, studied elegance, enhanced by such small decorative details as the thin, linear grille and...sharp, prow-like leading edges." These were wonderful kudos for a car that narrowly escaped oblivion.

Tjaarda's contribution to the design of the first Lincoln-Zephyr cannot be overlooked because his overall design caught Edsel's attention. However, Edsel's and Gregorie's contributions must also be recognized: Edsel for seeing the potential and foibles of the design, and Gregorie for developing a front-end treatment that satisfied his patron's discerning eye. The Zephyr program was the spark that ignited the creative synergism that eventually developed between Edsel and Gregorie. As patron and designer began working together more closely, Gregorie's ability to "see" what Edsel was visualizing manifested itself. How else could Gregorie explain how he was able to transform so successfully Edsel's nonsensical idea of adding a pointy front end to Tjaarda's bulbous design? The answer, says Gregorie, is because he and Edsel began to "think as one."

CHAPTER NINE

THE CONTINENTAL CAR III
"I Like Right-Hand-Drive Cars"

Gregorie had always thought that the Ford chassis was too high and too short to really accommodate stylish designs. He says that the Fords he designed during the 1930s were attractive but more "spunky" than stylish. For Gregorie to have made stylish Fords, he would have needed an entirely new chassis. However, at the time, only one chassis was available, and Gregorie had to do the best he could with that. Plus, convincing Henry Ford of the need for a longer, lower Ford chassis might be difficult, because Henry already thought the Ford car was too long and too low as it was.

Edsel Ford never gave Gregorie a directive to develop a simpler chassis than the one used on their latest continental car, but Gregorie knew it was an obstacle that needed to be overcome if he ever expected to see Edsel's sports-car idea go into production. He pondered the problem at great length and eventually developed an idea that was nothing short of genius. Perhaps it was his conservative nature or his thorough understanding of Henry's penchant for simplicity, but Gregorie designed and built a chassis from standard 1934 Ford parts that was significantly lower and longer than the standard version. To reduce the height of the chassis, Gregorie, with the help of Harold "Robby" Robinson, the Lincoln plant manager, took a standard Ford chassis and cut it in half directly in front of the rear axle kick-up. Then they took the rear half, turned it upside down, and welded the pieces together again so that the rear frame kick-up lay under the rear axle instead of over it. In other words, the rear part of the frame was now underslung, which lowered it a full six inches. In the front, Gregorie left the front cross member and spring in their original location and orientation, but he moved the axle forward by an ingenious spring-to-axle mounting mechanism that he designed himself. (Figures 9-1 and 9-2) By utilizing these brackets, which added only fifteen pounds to the chassis, Gregorie retained the stock Ford axle, spring, cross member, and steering shaft, and increased the wheelbase a full ten inches!

Gregorie's innovations were effective, for during cursory test drives, the chassis exhibited "excellent road-handling characteristics." However, Gregorie became concerned over a comment

Figure 9-1. The right front side view (top) and underside view (bottom) of Gregorie's special chassis show the ingenious method he used to lower and lengthen a standard Ford chassis to use in his continental cars. The spring and cross member remain in their original positions, but the axle is moved forward ten inches. (On a production Ford, the axle would be directly above or slightly behind the axle.) Note also that this is a right-hand-drive chassis; the bottom end of the steering column, located on the right side of the frame, can be seen above the starter and behind the exhaust pipe. (Photos from the Collections of Henry Ford Museum & Greenfield Village)

Figure 9-2. Through this photo taken at the Lincoln plant, the complete version of Gregorie's extended wheelbase Ford chassis can be seen. Note the position of the front wheels relative to the radiator shell, and compare it to the production 1936 Ford (Chapter 11, Figure 11-4). This is a production chassis, which has been unaltered except for the relocation of the front wheels using Gregorie's ingenious spring-to-axle mounting brackets. Moving the front wheels to this location would have given the 1935, 1936, and subsequent Ford models a longer front fender profile and a better-proportioned appearance. Also, note that the steering column is on the right side. (Photo from the Collections of Henry Ford Museum & Greenfield Village)

made by Larry Sheldrick after seeing the makeshift chassis. "You don't expect risking Mr. Ford's life in that contraption, do you?" he sniped.

The experimental chassis was well made, but Sheldrick's offhanded remark made Gregorie realize that it might be a good idea to evaluate the chassis under typical highway conditions; thus, he had another one built for that purpose. The second chassis incorporated Gregorie's unique front suspension, but because it was built on the all-new 1935 Ford chassis, it did not require any cutting or welding in the rear to lower it. The 1935 chassis, unlike the chassis of the 1934 car, had a higher rear axle kick-up, which enabled it to be lowered without major alteration, making the conversion easier. For no reason other than his predisposition for foreign cars, Gregorie had the chassis equipped with right-hand drive, using the parts he obtained from the Windsor plant of Ford of Canada, located across the river from Dearborn. "I like right-hand drive cars," explains Gregorie, "and had this chassis built up like that. It had no significance other than it was a personal preference. Nobody had any particular objection to it." Building the car with right-hand drive or left-hand drive would not affect the testing of the chassis, which was the purpose of the entire project. However, building the car with right-hand drive proved to be uncannily prophetic.

A bare chassis could not be tested on the open highway (although the testing of prototypes on state highways was a typical practice at Ford). Therefore, Gregorie quickly equipped the chassis with improvised body work made from a concoction of 1934 and 1935 production parts. (Figures 9-3 and 9-4) Although not intended for evaluation or consideration as part of the continental car project (it was constructed only to test the chassis), the improvised roadster appeared handsome and sporty nonetheless. "It started out with a 1934 roadster cowl, just something to support the steering column on while I drove the hell out of it around the test track and back country roads," explains Gregorie. The front doors were cut-down Ford roadster doors. The grille, front bumper, and headlamps were all production 1935 Ford parts. "The rest of it," says Gregorie, "was just sheet metal improvised by the aircraft boys. They were real pros at working sheet metal. They built a special hood, and then they cobbled up some front fenders from a couple of Tri-Motor fenders. The rear fenders were standard '34 Ford fenders, and the body structure around the tail end was just good aircraft boys handling sheet metal." Although "there was no design suggestion or intention meant at all in the body," Gregorie gave the simple body a touch of panache by adding two large accessory driving lights directly above the front bumper—lights that he had removed from his 1930 Stutz before selling it. He outfitted the right side of the car with polished aluminum disc wheels and the left side with wire wheels, to determine which gave a more pleasing appearance. "It's funny," recalls Gregorie, "but nobody ever noticed that!"

The liberties Gregorie took in building these experimental cars—the boattail designs, the special frames, and the right-hand drive—illustrate how much leeway Edsel gave Gregorie in the creative process. Edsel would always set the direction, not quite sure where it would lead, but he had confidence in Gregorie to develop something appropriate.

By the time Gregorie had completed building his experimental chassis and outfitted it with the improvised body, Edsel was prepared to approve the manufacture of a Ford-built continental car if the chassis proved durable. For that reason, he instructed Gregorie to personally drive the makeshift car back East, not only to get the requisite highway test miles on it, but to have a number of custom-body manufacturers bid on making a sporty body for his long, low chassis.

Although it was a business trip, Gregorie and his close friend Walter Kruke, who went with him, made the best of it, driving the right-hand-drive car to the body manufacturers in the dead of winter. "We didn't have a heater in the car," recalls Gregorie, "and we just had side curtains on it. The snow was flying around. It was a rough trip. If we hadn't had plenty of applejack with us, I would never have made it." Gregorie drove from the right side, while

Kruke sat on the left. The two jesters had fun mocking the rural folks as they drove the little roadster through Ohio and Pennsylvania. As they drove through the various towns along the way, Kruke would hoot and holler at the people walking along the street to get their attention. As they looked to see who was yelling, he would wave at them with both hands, as Gregorie sped down the street! The small-town residents had never seen a right-hand-drive car and were amazed at how the car could be driven through town with the driver's hands off the wheel! No matter how many times they did this prank along the route, Gregorie and Kruke always received a reaction of bewilderment from the unsuspecting onlookers.

In those days, Ford Motor Company did not build any of its own bodies. It built the chassis, fenders, hoods, and grilles, and outside body companies such as Briggs, Budd, and Murray built the bodies. If everything fell into place, Edsel and Gregorie would build their new sports car using the same technique—that is, have Gregorie's chassis built at Ford and the bodies built by an outside body company that was yet to be determined.

Unfortunately, their plan was quickly derailed by unforeseen complications. First, after taking his latest prototype to several body builders, some in Detroit and some on the East Coast, Gregorie was unable to find one interested in the project. The volume was either too large for the custom-body houses or too small for the volume suppliers. (Brewster expressed some interest in building the bodies, but only if Edsel paid for the construction of a new plant in which to build them. This was unacceptable to Edsel.)

Figure 9-3. Built in late 1934 using Gregorie's modified Ford chassis and some 1934 and 1935 production Ford parts, this last of the continental-type cars was built only to test the durability of Gregorie's innovative chassis and to give the custom-body houses an idea of what size and length of body Edsel was contemplating. At this time, Edsel was ready to give approval to place a car of this size into production. The two driving lights attached to the bumper were taken from of Gregorie's Stutz roadster. "They didn't work," says Gregorie, "but they gave the car that continental look." (Photo from the Collections of Henry Ford Museum & Greenfield Village)

More importantly, Ford Motor Company's production staff—Charles Sorensen, Pete Martin, and Henry Ford—simply did not like the idea. They felt that the production system of the company was too regimented to add another chassis to its high-volume production line.

However, the real reason was that Henry Ford simply did not want one of Edsel's newfangled automobiles to adulterate his functionally oriented car lines. Henry thought he had already allowed Edsel to place too much emphasis on styling, and he was not going to permit Edsel to produce a Ford as exotic as these continental cars. Henry did not care much about what Edsel did with the Lincolns and Lincoln-Zephyrs because both were manufactured at the Lincoln plant, which had been Edsel's domain for a long time. However, when Edsel approached him to add a sports-type car to the Ford line, Henry put his foot down.

This latest request also confirmed in Henry's mind that Edsel was being influenced too much by Gregorie. In fact, Henry often blamed Gregorie for planting new ideas in Edsel's head, thinking that if left to his own devices, Edsel would never have departed from his inbred practicality to start pursuing these sports cars, without being coaxed by somebody who had grandiose plans.

Occasionally, Henry would wander into Gregorie's design studio looking for Edsel. Wishing to be polite, Gregorie would ask the old man if he could show him a few things on which the design staff was working. Henry was not at all interested, and he replied by throwing up his hands in disgust or bewilderment and simply walking away, saying, "That's between you and Edsel! That's between you and Edsel!" As Ross Cousins recalls, "It was never a comfortable experience for either Henry or Gregorie when Henry visited the design department."

Figure 9-4. Gregorie poses with the third continental car that he designed for Edsel Ford. This view clearly shows the long, low effect Gregorie achieved with this design. Edsel gave this car to Gregorie after the project was abandoned. "I kept it for a couple of years," says Gregorie, "and I made several trips in it back East. I finally sold it to a friend of mine on Long Island for five hundred dollars. He couldn't wait to get that car!" Of all the cars he had specially built, it is this car that Gregorie would like to have today. (Photo courtesy of E.T. Gregorie)

Edsel might have been able to push the project through if one of the production men supported his idea, but none of them did. There was no way his father would be supportive because, as we have seen, Henry had absolutely no interest in design. Pete Martin, chief manufacturing executive at Ford, would not have been supportive because his authority had all but disappeared by this time. On the other hand, Charles Sorensen, the most powerful member of the Ford manufacturing troika, had some understanding of design as well as the authority and ability to persuade the others to be supportive. However, Sorensen "had a knack of taking the opposite view of anything that Edsel wanted," Gregorie explains. "I can see Charley Sorensen's lip curl now at Edsel's recommendation of building a sports car. He was, in essence, saying, 'The hell with production! Let's build a cute sports car!' But anytime the sports car idea came up with the potential of interfering with high production, why, somebody is going to be irritated. So I can see why Sorensen didn't want to go along with Edsel."

Therefore, Sorensen sided with Henry Ford and his cronies, and the idea of producing a continental-type car was tabled before it had a reasonable chance of beginning, at least in the United States.

Sometime between the construction of the second continental car and the third continental car, Percival Perry visited Edsel in Dearborn. After one of their discussions (no doubt about automobile design on the continent), Edsel began wondering whether his sports car idea could be brought to life in England. After all, Edsel conceived of the idea in England initially, and he did have the Ford plant at Dagenham. Edsel mulled over the idea, as he normally would do, but he did not pursue it until after Gregorie and Kruke had returned from their trip to the East Coast.

Edsel had some photographs taken of Gregorie's third continental car and sent them to Perry. He asked his European manager "if this type car is anything that might be interesting to you as an auxiliary line for the V-8 motor, and that could be built by one of your leading dealers if you did not care to attempt it at Dagenham.

If you would be interested in going into this matter further, I shall be glad to send the car over to you with drawings of the chassis changes, and if necessary, the young man that built it. [That young man, of course, was Gregorie.] The car has been tested thoroughly and has stood up to every requirement of a vehicle of its type. The body is a makeshift job built out of phaeton parts, with a new rear compartment added, and has nothing to do with the job other than to show what can be designed for it."

A month later, Edsel received Perry's reply, and it was not encouraging. He told Edsel that he did not think there was "any remunerative business for the company in sports vehicles...." Perry was having production problems with Briggs as it was, and he could not envision adding another body type to its production when he was not receiving satisfactory product now. When Perry described the project to a number of his biggest Ford dealers, none was interested.

However, being the good company man that he was and knowing how interested Edsel was in the sports-car project, Perry sent the photos to Lieutenant Colonel John Moore-Brabazon, an English World War I flying ace and champion race car driver. It is unclear why Perry chose Moore-Brabazon, but obviously they either knew each other well or Perry knew of Moore-Brabazon's desire to enter the automobile business. Whatever the connection, Moore-Brabazon was intrigued with Gregorie's sports car design and quickly made Perry an offer, asking for an exclusive license to market the special chassis. Perry sent Moore-Brabazon's letter to Edsel, asking if this arrangement would meet with his approval.

Edsel must have been overjoyed, because he rapidly approved the proposal. He told Gregorie about the arrangement he had approved for Perry and asked Gregorie to build another chassis—also with right-hand-drive—to send to England. "I have placed in process at the Airport," Edsel wrote Perry in September 1935, "a chassis which will be prepared and forwarded to you, together with detailed drawings, as soon as possible."

Perry received the chassis and drawings in October, and when he showed them to Moore-Brabazon, he realized he had made a good decision. "I must say," Moore-Brabazon wrote Edsel, "that the conversion is very ingenious....I think something can be done with a model like this....I am anxious to...gamble on it...."

While Moore-Brabazon was awaiting Edsel's reply, he contacted Richard and Allan Jensen, owners of Jensen Motors, Ltd., a body manufacturer in West Bromwich, a small town northwest of London. He also contacted Harold Kahn, a long-time business associate. Moore-Brabazon was well acquainted with the work of the Jensen brothers and hoped he could arrange a business venture with them. Although enthusiastic about the sports car project, he did not want to shoulder the risk entirely alone and thus asked Kahn to join him. Shortly thereafter, the two men formed a company called M.B.K. Motors to market the special Ford chassis.

Raised on the outskirts of London, Richard and Allan Jensen were two brothers bent on producing their own motorcar. Their first foray into automobile manufacturing was revamping a 1925 Austin that was given to them by their parents as a joint birthday gift in 1928. They designed a new body for the little car, lowered it three inches, and souped-up its engine. By the time they were finished, the car was quite an eye-catcher. In fact, the chief engineer of the Standard Motor Company was so impressed by the design when he saw the car running through the streets of Warwickshire that he asked the young men to modify a Standard chassis and equip it with a body of their design.

In 1931, the Jensen brothers joined the custom-body firm of W.S. Smith & Sons in West Bromwich. In 1934, they acquired control of the company and changed its name to Jensen Motors, Ltd. Initially, the brothers continued the old company's tradition of building custom bodies for expensive makes such as Rolls-Royce and Delage, but soon they increased the production within the company by offering custom-made bodies for all types and sizes of motorcars, including the little Wolseley Hornet and the Ford Model Y. Their reputation for producing attractive and sporty bodies grew, and with it grew their desire to build a car of their own. By late 1934, they had developed a design that would bear the Jensen name.

As with many manufacturers before them, the Jensens planned to build their new car by utilizing production components from other manufacturers. Their company would build the body, but they were in a quandary over what power plant to use and the expense of manufacturing a suitable chassis. When Lieutenant Colonel Moore-Brabazon entered their doors in 1935, he solved both problems for them.

After Moore-Brabazon gained possession of the special chassis, he showed it to the Jensen brothers. They were delighted with the way it performed, how it could be readily adapted to their body design, and with Moore-Brabazon's proposal.

By the summer of 1935, agreements among all parties had been reached. M.B.K. Motors would purchase power trains—V-8 engines, transmissions, and rear ends—from Ford in Dearborn, and the chassis components from Ford, Ltd. In turn, M.B.K. Motors would resell those items to Jensen Motors at cost. M.B.K. Motors would make its money on royalties paid by Jensen Motors for each car sold.

By the end of the year, the Jensens were selling sporty cars with the Jensen name. Similar to what Gregorie had done in his latest prototype, the Jensens designed and built the bodies for their new car utilizing a number of standard 1936 American Ford parts. They used the grille and front bumper as they were, but they lengthened the hood, running boards, and front fenders to give a more sweeping and graceful appearance—the same thing that Gregorie wanted to do on all American Fords. Interestingly, the Jensens added large headlights and touring lights to give their cars a more impressive appearance, something that Gregorie had done as a lark on his improvised roadster. (The accessory lights on Gregorie's car did not even work. "I added them just for looks," Gregorie says.)

The Jensens' new automobile instantly became a classic. (Figures 9-5 and 9-6) With a long hood, sweeping fenders, and a low profile, the two-door, four-passenger Jensen appeared sporty and nimble. Most were sold in the United Kingdom, but because of their elegant appearance and sporty performance, several made their way to the United States, most notably to Hollywood. In 1936, Clark Gable ordered one of the first Jensens—although he refused delivery after it arrived, finding the spectacular Duesenberg more enticing.

"The thing that amused me," says Gregorie, "was with all of Edsel's problems that he became so interested in this Jensen project. It struck a soft spot in his heart, because he liked that type of banter with someone developing a cute car. He was very sincere about it—he gave it his personal attention, far beyond its importance. In other words, there was no big money involved. He was just glad that he could be of some help to the Jensen brothers in getting their sports car off the ground. But it amazed me that Edsel Ford never brought one of those cars over to Dearborn. I wish he had. I wish he would have brought one over for himself and one for me. If we both would have had one, why, it would have been fine!"

The Jensen sports car also confirmed in Edsel's mind that he and Gregorie had been on the right course with their continental car program. He only wished that he could manufacture a similar car in the United States. Unfortunately, with the resistance he was receiving from his father, Edsel was forced to abandon his sports-car idea, at least for the moment.

Figure 9-5. This 1936 Jensen was built by Jensen Motors, Ltd., in England. Edsel Ford wanted to build an automobile similar to this for the American market. (Photo courtesy of Mike Lamm)

Figure 9-6. Edsel Ford, wearing the light-colored overcoat, visited the Jensen Motors factory in England in the summer of 1936. To Edsel's right is Richard Jensen, and to his left is Allan Jensen. The man at the far left is Percival Perry. The car is a 1936 White Lady, the Jensen brothers' first attempt at mass production. (Photo courtesy of E.T. Gregorie)

CHAPTER TEN

THE FORD DESIGN DEPARTMENT
"Anything the Boy Wants"

The months following the completion of the latest continental car was a slow time for Gregorie. The designers at Briggs were hard at work developing the next generation of Fords. Thus, the only thing that Gregorie could do was work on new ideas for Crecelius and Edsel, most of which never saw the production floor. However, that was about to change. In February 1935, Edsel took his annual winter vacation at his Florida home in Hobe Sound, and there he finally decided to establish a formal design department within Ford Motor Company.

Edsel had wanted his own design department for years, but his father's indifference to styling prevented that. However, with the success of the Model A, the 1932 Ford, and, more recently, Gregorie's designs of the Model Y and 1933 Ford, Edsel sensed the time was right to make the move. Gregorie's skills played no small part in Edsel's decision. Edsel had seen how well Gregorie could design automobiles to his liking and how forcefully Gregorie withstood the sniping and insults from the manufacturing men. Consequently, Edsel felt that Gregorie had the ability in all areas to direct a design department.

Nobody knows if Edsel discussed his decision with his father. If he did, he probably telephoned Henry from Florida, with his argument prepared in detail beforehand, and while his father was at home relaxing next to the fireplace. Edsel was not afraid to make decisions; he simply detested confrontation. Usually, if forced into a position to fight for his point of view, Edsel would walk away from the situation. He wanted the design department too much to set himself up for disaster by talking to his father while Sorensen and Martin were standing by Henry's side. It would have generated too much resistance for Edsel to overcome.

"As I look back on it now," explains Gregorie, "I think Edsel saw an opportunity—not being face to face with management up in Dearborn—to give an order over the telephone. I won't say he was a moral coward or anything like that. He just didn't want to get embroiled in a big discussion over the establishment of a design department. (If any dirty work was involved, he'd rather be out of town.) Nevertheless, he ordered it to be done without any face-to-face encounter with his father and the others."

It happened in the afternoon, shortly after Gregorie had returned from lunch. He was working on some design ideas at his desk when the telephone rang.

"Mr. Gregorie," he answered.

"Hello, Mr. Gregorie. This is Edsel Ford."

"Hello, Mr. Ford. What may I do for you?"

"Well, I've been thinking about our design activities and would like to ask you if you would like to head up your own design department?"

There was silence on the other end. Gregorie was flabbergasted.

"I've already talked it over with Charley Sorensen," continued Edsel, "and he's prepared to revamp the engineering department however you'd like it. How does that sound?"

Gregorie, in shock, hesitated a moment and then finally said, "I'd be overjoyed, Mr. Ford."

"Fine. Fine," replied Edsel. "Why don't you get with Charley as soon as you can and get things going?"

With that one phone call, Gregorie went from being a lowly "draftsman" in the eyes of Henry's cronies to a high-ranking Ford executive equal to that of Sorensen, Martin, or Galamb. Gregorie had been reporting directly to Edsel for approximately a year, but the others regarded him as Edsel's consultant rather than an executive. Now, Gregorie was assigned the responsibilities of managing a full department, one that would eventually become as important as manufacturing and engineering. Gregorie was only twenty-seven years old.

"It's strange how it happened that way," Gregorie recalls. "Edsel gave the order, and things had to happen. He was, after all, president of the company. He wanted things moved in that direction, and he didn't want any more dealings with Briggs as far as design was concerned. Everything was closed off at that point."

Interestingly, Gregorie did not have to search for Sorensen. "Charley Sorensen got his orders from the bossman," Gregorie explains, "and he came to me and told me that he was at my beck and call to establish the design department with full steam ahead, no strings attached. And, oh, he was sweet as honey! He was usually hard-boiled and then some. But, boy, syrup couldn't have poured out of his mouth any sweeter than he was! 'Anything the boy wants,' he told me. 'Anything the boy wants. Why, you just tell me, and I'll fix it up for you.' By the time Edsel came back a month later, why, we were going full bore."

Edsel made Gregorie the first chief designer at Ford because of Gregorie's extraordinary talent and also because of Gregorie's integrity and fortitude. "He knew I could handle my end if he backed me up," Gregorie says. "I couldn't have managed things without his backing. But he needed me, too. We had an unspoken working arrangement between us. I expected his backing where it was needed, and if he needed my backing, why, I'd try to go to bat for him. Many times, I would get into a heated discussion about this issue or that issue with the engineers, and Edsel would stand back and nod his head, and things happened. He apparently went to his father and just told him that's the way it had to run, that that was the new policy. The old man could see that styling was a big issue. So that left the chassis people to their own p's and q's. Edsel very subtly let these people know that I wasn't to be messed with. They were to work with me or else. And Edsel gave the orders that all designs had to be cleared through my office before production could begin. That was all it took to get the respect that I needed to establish the department and get respect for the way it was operated. So that was the break-off point. Oh, I had to stand up for the department. I took a chance, and it worked out."

Meanwhile, Sorensen ran into Walter Briggs in Palm Beach and told him that Edsel had decided to establish his own in-house design department. The company would continue to buy bodies from Briggs, but as far as body *design* was concerned, Briggs's services would no longer be needed.

The next day, Briggs called Ralph Roberts, his chief designer, and asked him how much work he had done on the Ford account. "Well, we have quite a few design concepts and several wood models completed and several more well underway," replied Roberts.

"Well, gather up whatever you have and take it all over to Dearborn," ordered Briggs. "We have just lost the Ford account."

"It was just that quick," explains Roberts. "By the time we had collected all the materials and models, as Walter had directed, we had about six truckloads of material to drop off at Ford."

Sorensen could not have cared less about design, but he may have been relieved that Edsel was not pulling out of Briggs altogether. Although it was never proven, some individuals at Ford and at Briggs thought Sorensen was receiving some "cushmo" from Briggs on the body orders Ford placed with the company. Like Gregorie, Sorensen was a boatman, "but he never went sailing," Gregorie says. "He went steaming! I had a 46-footer. But Sorensen's boat was a 146-footer, powered by twin 400-horse diesel engines! It was one of the biggest yachts in Detroit. Of course, a boat like that was relatively cheap in 1931. I imagine it didn't cost him a hundred and fifty thousand dollars!" Sorensen probably could have afforded such a magnificent boat on his salary, but it would have been much easier for him if he were receiving supplementary income from Briggs. In any case, that would not have lasted much longer either, because Edsel also intended to bring body manufacturing into the Ford fold and did so during the next few years. Pulling one body type away from Briggs at a time, Ford was building all its own bodies by the early 1940s.

Edsel did not have any animus toward Briggs or its designers. In fact, he thought the company did a commendable job for him. He pulled out, says Gregorie, "because he wanted the privacy to make his own design decisions in his own quarters. Over at Briggs, his decisions were exposed to a lot of other people who were heavily involved with design work for other companies, not just Ford. Edsel was glad to pull away from Briggs and have his own design establishment and to find someone who would carry out his desires. That is the way he wanted it. He was a lot more comfortable that way."

Sorensen and Gregorie decided to set up the new design department in the southwest corner of the engineering laboratory, which is an indication that Edsel must have received permission from his father before making the momentous decision. (Figure 10-1) The southwest corner was where Henry used to conduct his old-fashioned dances, forcing his executives to learn polkas, cotillions, and quadrilles. "To some of us," said one of the old-timers, "that was sort of sacrilege and a sign of changing times, when styling could take over Mr. Ford's dance floor." Although Henry Ford had discontinued his folk dances, he kept a gypsy orchestra of about six musicians to play there a few times each week. They would play for one hour or so, starting at three o'clock in the afternoon, over the wall from the design department. "Every now and then," Gregorie says, "one of those musicians would hit the cymbals or the drum just right...BONG!...nearly startling me and my boys off our chairs!"

Nevertheless, the location was ideal. With skylights for natural light, a beautiful and expansive teak floor, and floor-to-ceiling windows overlooking a manmade pond, it was an environment perfectly suited for creating.

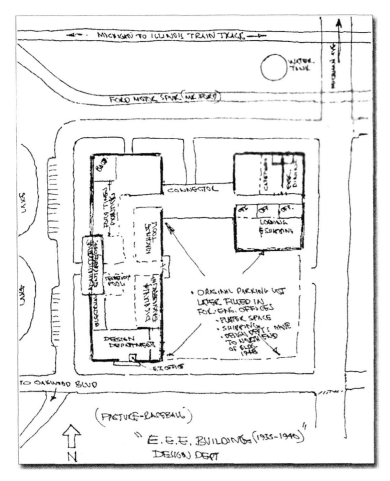

Figure 10-1. This sketch of the layout of the Ford Motor Company Engine & Electrical Engineering Building and grounds shows the location of the newly formed design department. In winter, Gregorie's "boys" would ice skate on the frozen ponds located in front of the complex. (Sketch courtesy of John Najjar)

The design department was always kept immaculately clean. No matter how much clay was "thrown" during the day, a cleaning crew swept and polished the floors at night. "The floor was so slippery," recalls John Najjar, a young designer at the time, "that to wear leather-healed shoes into that place was like taking your life into your own hands! I remember Emmett O'Rear, an apprentice like me, having some fun with Jimmy Mearns one day. Jim was a wood craftsman and was kneeling down, making a wood template from a clay model. Emmett, full of humor, took a rubber stamp title block, crept up behind Jim, and pressed the block onto the Jim's bald head. Emmett was wearing leather-soled shoes and had quite a time escaping Jim's wrath. It was funny seeing Emmett bounding around the clay models with Jim in full pursuit!" (Figure 10-2)

No permanent alterations were made to Henry Ford's sacred area, however. The new design department was merely separated from the engineering department with partitions made of yellow oak and frosted glass, having double doors at regular intervals to enable large tables and models to be moved in or out of the design department as needed. (Figures 10-3 and 10-4) The model shop also was separated from the design department with these partitioned walls. Curiously, directly above the double-door entrance to the shop was a large stuffed black crow hanging from a wire from the ceiling! Nobody seems to remember why it was there, but the bird was probably hung there at the order of Henry Ford himself during the dancing days. Nobody had the nerve to remove it for fear of reprisal from old Henry. "It had a wing span of about three feet," recalls John Hay, a craftsman in the shop. "They were strict about keeping the entire area clean, and every once in a while, my boss, Jimmy Lynch, had me get a ladder and climb up and dust that doggone bird off!"

Outside the design department, between the engineering laboratory and Oakwood Boulevard, were two manmade ponds that formed a water reserve in case of fire. The ponds also gave Gregorie's men opportunities to ice skate in winter and sail model boats in summer. Eugene "Bud" Adams, another one of the early designers, vividly recalled arriving early at work in the winter to ice skate on the frozen pond while the sun rose. Today, the old engineering lab, which remains in use, is practically surrounded

Figure 10-2. Jimmy Mearns, a wood modeler in the Ford design department, puts finishing touches on a wooden armature, circa 1945. Upon this well-made foundation, hundreds of pounds of clay would be applied by the clay modelers and sculpted into a new Ford, Mercury, or Lincoln design. In the background at the right is the rear end of a postwar station wagon. (Photo courtesy of Ford Motor Company)

Figure 10-3. In this earliest scene of the Ford Motor Company newly formed design department, circa 1937, E.T. Gregorie is standing at the left, next to the last drafting board, with a ruler in his hand. The other designers are (from front to back): John Walter, Gregorie's design supervisor in charge of steering wheels and instrument panels; Eugene "Bud" Adams; Emmett O'Rear; Bob Thomas; and Placid "Benny" Barbera. The man standing between Gregorie and Barbera is unidentified. The model in front of Walter's desk is of a bus to transport visitors, especially dealers, between the Ford Rotunda visitors center, the Rouge plant, and the train station in downtown Detroit. This bus design was never intended for commercial purposes and was never built. (Photo courtesy of Lorin Sorensen)

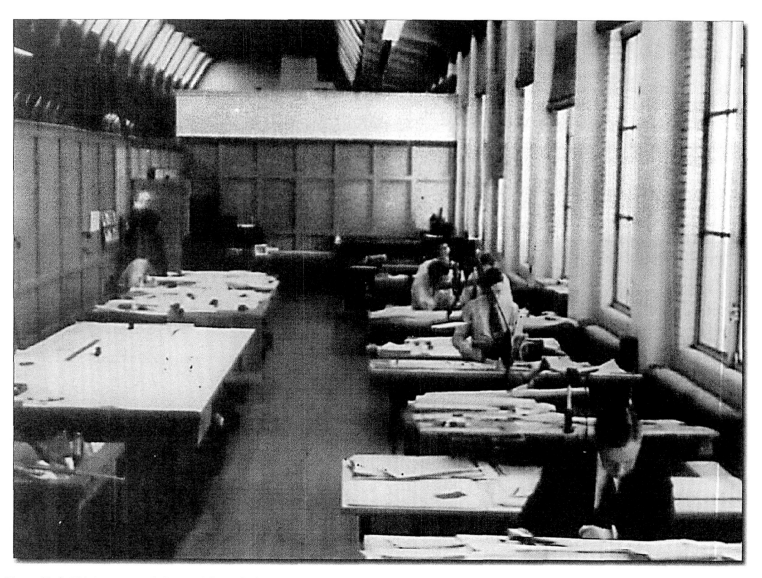

Figure 10-4. This is an expanded view of the early design department at Ford Motor Company. The design department was separated from the engineering department with oak paneling as shown, and as the department grew, the partitions were simply moved farther into the area that was once Henry Ford's dance floor. "When Eugene Adams and I arrived in the design department in 1937," explains Emmett O'Rear, "there were six drawing boards along the south wall as shown here. They were occupied by Walter Kruke, John Walter, and Willys Wagner. The other three boards were used by me, Bob Thomas, and Benny Barbera on a come-and-go basis." (Photo courtesy of E.T. Gregorie)

by a blacktopped parking lot; however, in those old days, the lab was surrounded by lush, green grass and majestic oak trees. Herds of sheep were brought there to keep the grass trimmed. When the weather was nice and the sheep were behind the gate, Gregorie would take the sack lunch his wife had packed for him and eat it sitting under a tree on the pristine grounds. Other times, he and one of his design managers would go into Dearborn and eat lunch at his favorite Greek restaurant.

Gregorie's office was centrally located along the south side of the engineering lab. (Figure 10-5) That wall formed one side of Gregorie's office. The other three walls were made of yellow oak, similar to the other partitions that composed the design department. However, instead of having frosted glass panels, Gregorie's office had clear glass panels, allowing him, in Johnny Hay's words, to "see the whole dang place sitting from his desk." Within his office, Gregorie had a desk, a drafting board on which he would do his design work, and a small table on which he would make small clay models. His office had two doors: one to the design department through the east partition, and one that opened directly to the parking lot through the south wall. Gregorie would park his car next to this south-facing door. When Edsel visited Gregorie after lunch, Edsel would walk down the long corridor from the executive dinning room, across the design department, and enter Gregorie's office through the east door. However, if Edsel wanted to visit his chief designer in private, he would park his car next to Gregorie's car, in a space marked "President," and enter through the outside entrance. Eventually, Gregorie had the periphery of his office surrounded by large moveable blackboards, the kind on which the designers made full-scale drawings, so that he and Edsel could discuss designs inside his office, without the whole department gawking at them through the windows.

Henry Ford's ironclad rule prohibited smoking in the design department, and no coffee or candy machines were located anywhere on the premises. "If you wanted to steal a smoke," explained Adams, "you went downstairs in the lavatory and were careful about it, because being caught smoking was cause for instant dismissal."

For years, only men worked in the design department. In fact, the reason it was called the design department as opposed to the styling department, explains Gregorie, "is that *styling* smacks of women's wear." As Adams recalls, "The only girls you ever saw were those from the secretarial pool on the second floor, either going to or returning from lunch or in the cafeteria itself." Women did not join Gregorie's staff until after World War II.

While Sorensen was managing the construction of the design department, Gregorie began recruiting men to staff it. The first man he hired was Dick Beneicke, a master clay modeler who was working at Briggs Manufacturing. (Figure 10-6) Beneicke "was an old, dumpy, lovable German who had learned his industrial-architectural sculpture trade in the old country," Adams said. "He came to the United States during the construction of the ornate movie palaces, whose interiors abounded with sculpture of all types. He subsequently joined Briggs Manufacturing, and from there was lured to Ford by Gregorie." While at Briggs, Beneicke always worked with a beautiful Meerschaum pipe hanging from his mouth. "I don't remember ever seeing him without his pipe," recalls Rhys Miller, who worked with Beneicke at Briggs for a short while. However, after Beneicke joined Ford, he too was forced to sneak a smoke in the basement lavatory with the other clandestine smokers. When Beneicke came to Ford, his close friend William M. Leverenz followed him there. Leverenz was also an architectural plasterer, and he had worked with Beneicke decorating movie palaces.

Beneicke taught all of Gregorie's young design apprentices how to construct clay models, and in doing so, he emphasized the difference between modeling and sculpting. "A modeled work piece has more warmth and love in it than a piece cut from a block of material and sculpted," he used to say. One day, Adams asked Beneicke how he knew when he had a good surface or a good line. After thinking for a moment, Beneicke replied, "It's kite simple," he said in his thick German accent. "You just add on vat isn't enuff und take off vat's too much!" He did make it appear

Figure 10-5. In this interior view of the design department at Ford, circa 1939, part of Gregorie's partitioned office can be seen at the right. The stairway in the center leads to the basement lavatories, where smokers in the department hid from Henry Ford's watchful eye. The man in the vest leaning over the drafting board in the lower right corner is Ed Martin, and the man in the vest facing the camera in the far back corner is Emmett O'Rear. The beautiful renderings in the foreground left were done by Gregorie's master illustrator, Ross Cousins. (Photo from the Collections of Henry Ford Museum & Greenfield Village)

Figure 10-6. Richard Beneicke was the first man hired by Gregorie to staff the design department at Ford. Beneicke is shown here in 1945, working on a quarter-size model of what appears to be a postwar Lincoln Continental. An expert clay modeler, Beneicke worked for Briggs Manufacturing before joining Ford. He was instrumental in converting Gregorie's two-dimensional drawings into three-dimensional clay models, and he helped train all the early design apprentices hired by Gregorie. Beneicke died of a heart attack approximately two years later, while getting ready for work. (Photo courtesy of Ford Motor Company)

easier than it was, but eventually many of Gregorie's apprentices learned to "rub da clay," as Beneicke used to say, as well as Beneicke could.

During those years, the modelers used Chavant clay, which was a gray-green, waxy substance. The clay softens when warmed in ovens or when kneaded by hand so that it can be molded, and it solidifies at room temperature so that it can be painted. "Clay shavings and virgin clay blocks were placed on aluminum trays and softened in ovens that were heated by light bulbs," explained Adams. "These ovens were designed and manufactured at Ford. When you wanted to get some clay, you'd go over to the oven and pull out a drawer and grab a glob of the stuff. Hopefully, it wasn't too hot so that it overheated your fingers, but you'd take that back to your model and proceed to rub it in place. While it was still warm, the modelers would smooth it and shape it with their hands and with spring steel scrapers made into different shapes and contours. Beneicke was the only one who initially had a set of these tools, but eventually all the clay modelers at Ford had copies of his tools made at the Rouge tool room."[a]

The number of men required to clay up a new model depended on time constraints, but usually it took six men eight to twelve weeks to take a design from a bare armature (the wooden framework used to support the clay) to a finished, full-size clay model. (Figure 10-7) Loading the clay was the most strenuous part of the job because the clay was heavy. Three men at a time would load the clay: one in the front of the car, one in the middle, and one in the rear. When they became tired, they would switch with the other three men, who would then continue loading the clay. The six men would alternate until all the clay was loaded on in rough form, which took approximately a week. "Loading the clay on was just not simply making a blob of clay," explains Najjar, "it is building the clay up to predetermined surfaces, usually set in with templates or some other points [taken] from our [modeling] bridge. You'd sometimes get clay six inches to nine inches deep on a model as it went from rough clay to finish."

Wayne Booth, a latecomer to Gregorie's staff, explains that loading the clay was critical to the final outcome of the model. (Figure 10-8) "If you didn't pack it tightly as you applied it," Booth explains, "the clay would collapse or crack as it hardened. It would also crack if the armature underneath was not strong enough or if the model was subsequently moved." As the clay became closer and closer to the desired surface, the designers would apply thin layers, one at a time, "layer after layer after layer." This was the painstaking part. After the model was roughed in, another eight to ten weeks were required to finish the clay model "so that it was ready for paint, or shown in its final clay surface, smoothed down with turpentine and aluminum foil applied to the moldings and door handles."

Bud Adams explained that "after working on a full-size clay model for months at a time, you get an 'educated' pair of hands. You can feel very slight irregularities on a roof panel, for example, which is fairly flat. It's a gradual curve, but after you had worked on a model for a while and learned how to scrape clay, you could [rub] your hand over it and feel any depressions in the surface. When you actually took a flexible straight edge and laid it in that area, you could just barely see light at the spot that felt low. It was actually less than a thickness of paper, so the hand developed a keen sensitivity."

This backbreaking and monotonous work prompted the young apprentices to devise ways to relieve tension and built-up energy. "One day, we were nearing the noontime lunch period," remembers Najjar, "and Dick Beneicke went out to lunch while we trainees ate out of lunch bags, so we had a lot of spare lunch time to use. We decided to pull a trick on Dick and proceeded to make footprints on a freshly loaded clay model. We moved the floating bridge aft as far as it would go and proceeded to run up the back of the model, over the roof, and down onto the hood and then to

---

[a] The Rouge plant is one of the Ford Motor Company manufacturing plants. It derived its name from its location on the River Rouge, approximately four miles away from the design department.

Figure 10-7. This photograph of the interior of the Ford design department in the fall of 1945 provides a good view of an armature, the massive wooden framework upon which hundreds of pounds of clay would be spread into the final form of a new Ford design. The men working on the armature are (left to right) John Duncan, Joseph Somogyi, and Jimmy Mearns, all wood modelers. The modeler in the background, who is working on the quarter-size clay model, is Michael Bogre, and the man to his left in the dark, sleeveless sweater is Ray Mecklenburg. The two men standing examining the full-size clay model are visitors, and the bald man to their right with his back to the camera is one-time custom-body designer, Tom Hibbard. (Photo courtesy of Ford Motor Company)

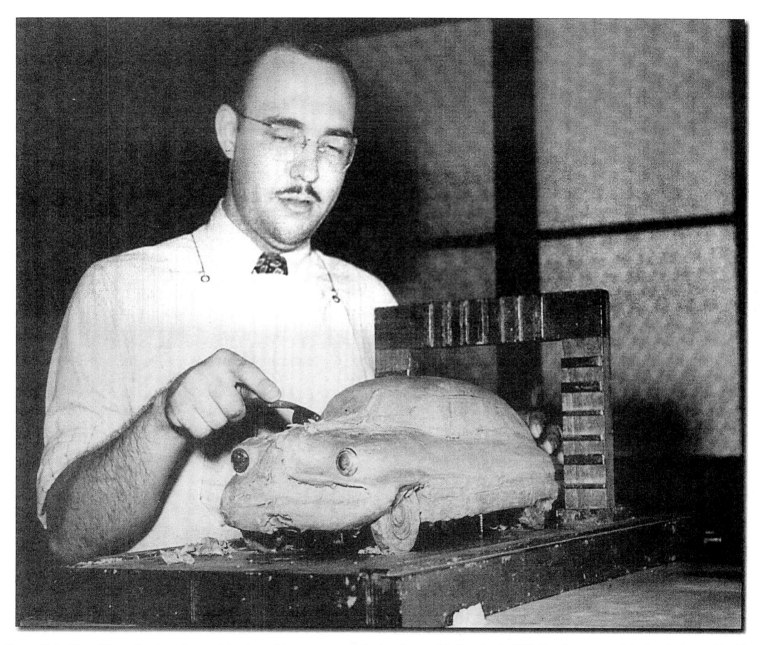

Figure 10-8. Here, Wayne Booth, a clay modeler, is working on a one-tenth-size clay model. Starting in 1945, Booth was one of the last designers hired by Gregorie. (Photo from the Collections of Henry Ford Museum & Greenfield Village)

the floor! It was quite a mess after a couple of passes, and by then we realized it might not be funny. Dick often wondered why we trainees were so busy cleaning up and loading fresh clay when he returned from lunch. It goes without saying, that E.T. Gregorie was not in the building when all this occurred."

"Those clay models were awfully heavy," explains Gregorie. "When we put a floor jack under one to put a tire on or something, why, you could hear the teak floor underneath squeak and squawk." For this reason, Dick Beneicke chastised the apprentices when they crawled under one of the models to make an adjustment or retrieve a dropped clay knife. "You stay oot from under dat!" he'd yell. "Dat, dat, dat ting might fall on you!" (Figure 10-9)

Beneicke's contribution to the early Ford design department cannot be overestimated. Described by one Ford designer at the time as "Michelangelo with a German accent," Beneicke's modeling expertise and eye for a good form brought Gregorie's sketches to life. "Dick Beneicke would walk around with Mr. Gregorie and Mr. Ford as they discussed the various models," explains Placid "Benny" Barbera, "and he would learn firsthand what Gregorie and Edsel were thinking. Beneicke then interpreted the ideas and sketches and all the verbal directions into three-dimensional form." Generally, in automobile design, the design sketch always appears better than the full-size clay model. However, at Ford, with Beneicke's expertise, the clay model—and subsequently, the produced automobile—always appeared better than the sketch.

Soon after Gregorie had brought Beneicke and Leverenz on board, several other men joined the fledgling design department, and Gregorie placed them in charge of four separate sections: (1) design, (2) body layout and engineering, (3) clay modeling, and (4) fabrication shop. Gregorie placed Beneicke in charge of clay modeling, with Leverenz as his assistant. Jimmy Lynch was in charge of the fabrication shop, with Bert Pugh as his assistant. Martin Regitko was in charge of body layout and engineering. Gregorie separated the design activities into four separate activities. In charge of grilles and exterior trim was Bruno Kolt; interiors, Walter Kruke; instrument panels and steering wheels, John Walter; and lighting, bumpers, and wheels, Willys Wagner. (Figure 10-10) Edward Martin assisted Gregorie in managing the design activities and was the design department coordinator with engineering.

All these men came from diverse backgrounds, but under Gregorie's direction they were melded into a formidable design department. For example, Walter Kruke, the tall and handsome Yale graduate who accompanied Gregorie on his East Coast jaunt in the last continental car, had wanted to become a professional musician while at college and had been a competitor of crooner Rudy Vallee. However, when Kruke's band failed, he took a job on one of Ford's Great Lakes freighters, and eventually designed interiors for Ford Tri-Motor airplanes at the Ford aircraft division. Gregorie knew that Kruke enjoyed working with fabrics, and that is why he put Kruke in charge of interior design.

Edward Martin was a body draftsman/engineer under Crecelius at the time Gregorie started the design department. "Ed was born in the Philippines," recalls Tucker Madawick. "He was part French and part Korean. In his late teens, he shipped on board as a junior radio operator, sailing the Pacific between San Francisco and Asian ports. This brought him to many cities on the West Coast and ultimately to working in one of the Ford assembly plants." Somehow, Martin landed in Dearborn, working for Crecelius.

Jimmy Lynch had worked at the Ford aircraft factory, and Gregorie realized Lynch's craftsmanship when Lynch had helped him construct Edsel's continental cars. When Gregorie started the design department, he made Lynch manager of the fabrication shop in the design department. Lynch and his men fabricated all types of prototype parts—from steering wheels to headlights to name plates.

Bruno Kolt originally came from Berlin, Germany, and had worked for a number of custom-body houses on the East Coast. As these companies began to fold during the Depression, Kolt came to Detroit looking for a job and was hired by Gregorie in 1935. When World War II started, Kolt wanted to return to Germany to "help

Figure 10-9. James Mearns, standing on a stepladder, assists young clay modelers in roughing out a full-size clay model, circa 1939. Tony Schuch is to his immediate left, and Robert Paulson is to his right. (Photo from the Collections of Henry Ford Museum & Greenfield Village)

Figure 10-10. Willys P. Wagner (left), one of the first men hired by Gregorie to work in the Ford design department, works with engineer Fred Mitchell on a full-size blackboard layout. Wagner was responsible for the design of taillights, trunk handles, and other body hardware. (Photo courtesy of Ford Motor Company)

the Father Land," but Gregorie talked him out of it. It was a good thing he did, not only for the obvious reason but also because "Bruno was one of the best designers Gregorie had," explains Barbera.

John Walter was a draftsman at Ford's Rouge plant before joining Gregorie's group, and Willys Wagner joined the department in 1935. Wagner had attended the University of Michigan where he received architectural training, and from there he had gone to work for Murray Corporation. He was hired by Gregorie to demonstrate automobile styling at the Ford Motor Company exhibit at the San Diego Exposition. When Wagner returned to Dearborn, he applied his talents to the design of taillights, license plate brackets, and other exterior trim.

In early 1937, Edsel had a practical idea that would not only help Gregorie staff his new department, but would also prevent Henry Ford's interference. As Emmett O'Rear explains, "It was common knowledge that Henry Ford never took any interest in anything Edsel did, even to the extent of literally destroying Edsel's projects. That was the reason that Edsel suggested bringing in Trade School boys to the design department." (Figure 10-11)

The Henry Ford Trade School was located on the second floor of the "B" Building at the Rouge plant. There the students were taught math, geometry, trigonometry, mechanical drawing, machine shop theory, and the operation of all types of metalworking machinery. "We had outstanding instructors," explains Barbera, one of the first trade-school recruits. "I learned how to run a lathe, shaper, vertical and horizontal mill, O.D. grinder and surface grinder, and spline mill." Many graduates went on to the Ford Apprentice School, where they were taught "advanced drawing, advanced math, and body surface layout." These skills proved invaluable for those lucky enough to be assigned to the design department.

"The Trade School was Henry's pet, and he supported anything where 'the boys' were involved," explains Emmett O'Rear. "E.T. Gregorie was not liked by the other executives. They felt that 'styling was a lot of nonsense.' Henry Ford tended to agree with that approach but tolerated it because of the boys. Therefore, if the design department was presented as a marvelous opportunity for students as they graduated, it would have a chance of survival." Thus, at Edsel's direction, Gregorie interviewed a number of Trade School graduates in the summer of 1937, and he recruited five of the most talented. Eugene "Bud" Adams, Placid "Benny" Barbera, Frank Beyer, and Emmett O'Rear (Figure 10-12) were recruited as apprentice designers, and Francis Zawacki was recruited as Gregorie's secretary.

John Najjar and Bob Thomas joined the department at this time also. As an indication that Edsel's scheme of hiring "Henry's boys" was working, Henry Ford himself sent them to see Gregorie. At the time, Najjar was a student at the Ford Apprentice School, and Thomas was an apprentice of Irving Bacon, the resident artist at Ford.

Najjar was working at a lathe one morning when he felt a tap on his shoulder. He turned around, and there stood Henry Ford. "Do you like this kind of work?" Henry asked.

"No, I don't," Najjar replied honestly, not knowing to this day why Henry Ford chose him from the other boys in the school. "I like to draw."

Henry then asked Najjar if he had any drawings to show him, and when Najjar replied that he did, Ford said, "Why don't you bring some in tomorrow so that I can look at them?"

The next morning, Henry Ford arrived as he promised, and Najjar showed him his drawings. Without saying much, Henry looked at them and then left.

Figure 10-11. The education and training received by students at the Henry Ford Trade School (shown here circa 1940) were second to none in the automobile industry. (Photo from the Collections of Henry Ford Museum & Greenfield Village)

Figure 10-12. Emmett O'Rear, one of the Trade School "boys" hired by Gregorie in 1938, is "rubbing da clay" on what appears to be a Lincoln-Zephyr. Henry Ford thought the designers, dressed in smocks, were "queer." (Photo from the Collections of Henry Ford Museum & Greenfield Village)

A few months later, Najjar's boss came to him and told him that the personnel department was seeking people who could draw to work in the design department. (Figures 10-13 and 10-14) Najjar went to see Gregorie, showed him the same drawings he had shown Henry Ford, and was hired. "My first job," explained Najjar, "was answering the phone, and having a sketch pad, and being told to draw instrument panel control knobs for John Walter."

In a roundabout way, Henry Ford also was responsible for Bob Thomas joining Gregorie's early design department staff. A native of Pennsylvania, Robert McGuffey Thomas was a relative of William Holmes McGuffey, the writer and publisher of *The McGuffey Readers*, the storybooks Henry Ford remembered so fondly from his childhood. When Thomas was in high school, Charley Newton, one of Henry Ford's antique hunters, visited the Thomas family in search of original *McGuffey Readers* to place in Ford's collection in Greenfield Village, Henry's recreation of old-time America. During his visit, Newton invited Thomas to come to Dearborn to meet Henry Ford because Thomas was related to McGuffey. Thomas met Ford a few weeks later, and the two spent most of the day together. Henry took Thomas on a personal tour of Greenfield Village, invited him to have lunch with him in the executive dining room, and after lunch, went for a walk in the engineering lab. Before they parted company, Henry offered Thomas a summer job in Greenfield Village. "Thus started a chain of events that predestined my future," recalls Thomas.

After Thomas finished high school, he went to work in 1936 on a full-time basis at Greenfield Village, working for Roy Shumann, who was responsible for dismantling historic buildings on their original sites and reconstructing them in Greenfield Village. Thomas' first job was to make engineering drawings of damaged or missing parts of the various machinery on display in the Henry Ford Museum. Shumann was so impressed with Thomas' drawings that he showed them to Henry, who in turn reassigned Thomas to work with Irving Bacon. Thomas worked for Bacon for approximately nine months, and by the end of that time they both realized that Thomas was not a painter. However, Bacon knew Thomas had

Figure 10-13. A young John Najjar poses outside the Ford design department in front of his 1936 Ford, shortly after starting with the company in 1938. Najjar was one of the first design apprentices hired by Gregorie, and he worked for the company for more than forty years. However, Najjar's most significant contributions came after Gregorie had left Ford: the 1956 Mercury Turnpike Cruiser, the 1961 Lincoln Continental, and the 1964 Mustang. (Photo courtesy of John Najjar)

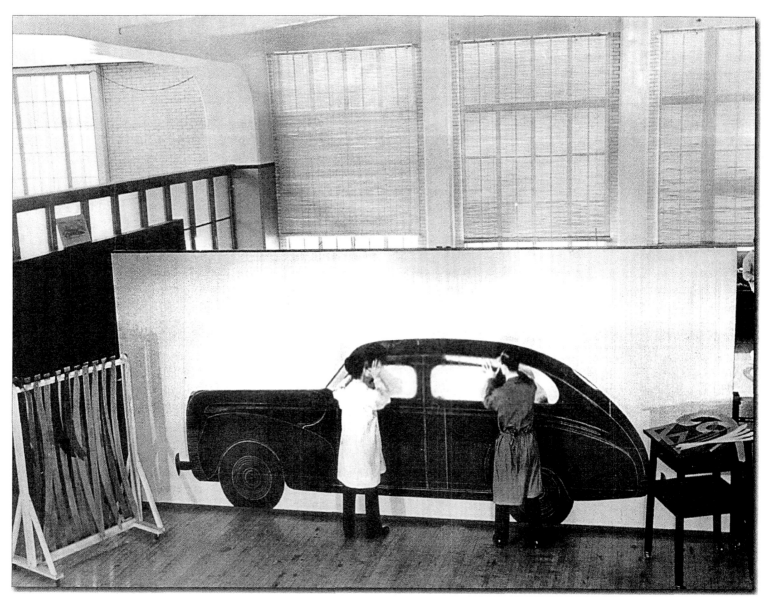

Figure 10-14. John Najjar (left) and Edward Martin (right) are making a full-size design drawing in the Ford design department, circa 1938. Note the rack of sweeps to the left of the drawing. Sweeps are long templates used for laying out different curvatures on full-size drawings. (Photo from the Collections of Henry Ford Museum & Greenfield Village)

artistic ability and took him to Gregorie to see if Gregorie could use Thomas in the design department. (Figures 10-15 and 10-16) "I showed [Gregorie] the work that I had been doing for Bacon, and he was impressed," recalls Thomas. "This was on a Friday, and he asked me to do some car drawings and bring them in on Monday. Whew! What a busy weekend. I made two watercolor car renderings and took them in on Monday. They weren't the greatest, but they were good enough for Gregorie to offer me a job." That was in the fall of 1937.

Soon after Thomas joined the design department, John Hay was assigned to work with Jimmy Lynch in the fabrication shop. Hay landed his job by submitting a fifteen-inch-long model of a 1937 Ford phaeton that he had made from tin cans and other scrap metal. It was a marvelous piece of workmanship. (Figure 10-17) It had a separate frame, transverse springs in the front and back, a banjo steering wheel, and working doors, hood, and lights. When Gregorie saw it, he immediately hired Hay for the fabrication shop. "Most everything we made out of wood: door handles, hood ornaments, moldings, display bucks, and instrument panels," explains Hay. "I remember making a molding that was tapered on each side and went straight down the middle of the hood and ended up with a hood ornament at the end. It was long and straight and beautiful."

No expense was spared for materials used in the design department. For example, the armatures for the clay models could have been made from any grade of lumber, even scrap lumber. "But not at Ford Motor Company," recalls Najjar. "They had a habit of ordering in number one clear pine, and that meant no knots in the wood!"

The last design apprentices recruited during this formative period were Tucker Madawick and Duncan McRae. In 1935, Madawick entered the Pratt Institute in New York. Pratt was the first university in the country to offer a degree in industrial design, and in 1938 Madawick was in the first graduating class. He was hired immediately by Ford Motor Company to work at its exhibit at the New York World's Fair. "Ford had a styling exhibit in the Rotunda Building set up at the fair," explains Madawick, "and I was hired to man one of those exhibits." When the fair ended in the fall, Madawick was transferred to Dearborn and became a member of Gregorie's staff. (Figure 10-18)

Duncan McRae also joined the Ford design department around this time. He had no formal design training as Madawick did, but he loved cars. He simply "knocked on the door," says Madawick, and Gregorie hired him. (Figure 10-19)

Ross Cousins was the last member of this initial group, but he was an illustrator rather than a designer. (Figure 10-20) Cousins learned his trade from his father, who was a commercial artist, and by taking art classes at Cass Technical Institute in Detroit. He graduated in 1935, and his first job was working at a print shop that did a lot of automotive work. From there, Cousins went to work for another printing company and eventually worked at Chrysler. While there, he heard about a job at Ford and applied for it. Gregorie hired Cousins to assist the advertising agencies in developing ads. "Ross could make nice drawings and show the car in front of a butcher shop, or a gal stepping out there and the guy putting groceries in the back and all that kind of stuff," says Gregorie. Many of the Ford advertisements showing illustrated scenes during the late 1930s and early 1940s were done by Ross Cousins.

"Those boys were eighteen, nineteen years old when they came to work for me," recalls Gregorie. "I was only in my twenties." He still refers to them as "the boys," although they are all now well into their seventies.

In addition to this diverse group of neophytes and experienced draftsmen, Gregorie also had several seasoned automobile designers on his staff. He quickly notes that he never solicited them; they all came to him looking for a job. (Gregorie did not want to give old Henry Ford the impression that he was establishing a professional organization!) Martin Regitko was the first of these

Figure 10-15. This photograph is an interior view of the Ford design department in 1939. At the far left, working on a design buck, is Bob Thomas, the designer who later would be instrumental in designing the 1956 and 1961 Continentals. Standing in front of the 1940 Ford Deluxe in the foreground is Emmett O'Rear, wearing the vest, and to his right is Bruno Kolt. The man standing left of center in the white smock is Ross Cousins, and he is showing one of his automobile renderings to Frank Francis, Gregorie's office manager. The man in the far background, wearing a white shirt and dark tie and leaning on the drafting table, is Martin Regitko. (Photo from the Collections of Henry Ford Museum & Greenfield Village)

Figure 10-16. This photograph shows the Ford design department, circa 1943. In the far left corner, slouched over the drafting board, is John Najjar. At the right, working on a quarter-size model of a postwar truck, is Cesare Testaguzza. (Photo from the Collections of Henry Ford Museum & Greenfield Village)

Figure 10-17. John Hay, one of Gregorie's early recruits in the design department, points out the features of his exquisitely detailed model of a 1937 Ford phaeton. Hay made this model from scrap pieces of tin cans and other metal pieces. Hay worked for the Ford Motor Company for forty years. (Photo from the Collections of Henry Ford Museum & Greenfield Village)

Figure 10-18. Tucker Madawick, one of the last design apprentices hired by Gregorie to join his fledgling design department, stands next to his new 1941 Ford convertible. Madawick's design career at Ford ended when World War II started. After the war, he helped design the Tucker automobile, working for J. Gordon Lippincott, and the 1953 line of Studebakers, working for Raymond Loewy. His design career culminated as design vice president of RCA. (Photo courtesy of Tucker Madawick)

men. Regitko came from Willoughby & Company, custom-body builders in Utica, New York. Regitko's forte was laying out the body structure, from internal door pillars and structural members around window openings, to exterior body surfaces. In the late 1930s, Hermann C. Brunn joined the Ford design department. Brunn had learned the art of custom-body design from his father, the founder of Brunn & Company, a notable custom-body builder in Buffalo, New York. From the custom-body house of Waterhouse & Company of Webster, Massachusetts, came Charlie Waterhouse, son of the founder of the firm. John Dobben, formerly a designer and later chief engineer for Judkins Company, and George Tasman, body designer for Locke and later LeBaron-Detroit, also were members of Gregorie's staff. Edsel Ford had purchased thousands of bodies from these firms over the years and no doubt had known all these men well before they joined Ford; however, it is unlikely that he encouraged any of them to move to Dearborn. After World War II, Thomas Hibbard, a former partner in the famous LeBaron body company in the early 1920s, joined Gregorie's staff from General Motors. "I'm not quite sure how Tom came to Ford," says Gregorie, "but he was working at the Rouge [plant], helping develop the Jeep. He came over to my department shortly after that."

These men were body draftsmen for Gregorie. "They had no great knowledge of production bodies," he says, "but they were very knowledgeable people when it came to drafting standards and body construction. They had a good foundation in the business, and they didn't need re-education. We were fortunate to get them."

Soon after World War II, the first women joined Gregorie's design staff. "I don't know what brought it on," explains John Najjar, "but we ended up with about six women in the department somewhere in the period of 1945 and 1947. It was very surprising—we hadn't heard of women being in the design business up to

Figure 10-19. Duncan McRae, one of the early design apprentices, sketching some ideas for the 1941 Ford. After leaving Ford in 1942, McRae worked for Chrysler, then for Kaiser-Frazer, Studebaker, and Curtiss-Wright, ending his career at Ford. (Photo from the Collections of Henry Ford Museum & Greenfield Village)

Figure 10-20. Ross Cousins, Gregorie's master illustrator, is sitting on the fender of his new 1946 Ford while painting one of his masterful renderings. Gregorie hired Cousins to help in producing design renderings and print advertisements. (Photo courtesy of Ross Cousins)

then." The first woman in the department was artist Leota Carroll. (Figure 10-21) She worked for Ross Cousins, who taught her how to make pastel renderings of automobiles. "Leota was a very attractive young woman," recalls Cousins. "The whole department lost about an hour's worth of production whenever she got up and walked down the aisle. She sat directly behind me, and I'll tell you, I dropped my pencil on the floor quite often!" Doris Dickason, a former art teacher, added a woman's touch to the design of interiors and details such as door handles and ashtrays. Elre Campbell, another interior and detail designer, owned a lampshade company before the war, designing one-of-a-kind lighting fixtures for theaters and offices. When World War II started, she got a job at the Willow Run bomber plant of Ford Motor Company, illustrating instruction manuals for production operations. Letha Allan, Beth O'Roke, and Florence Peterson, all of whom had similar positions as the others, also joined Gregorie's staff during this period.

By 1938, Gregorie had a staff of approximately twenty-five men, all earning sixty-five to seventy-five cents per hour. His staff eventually grew to approximately one hundred and twenty-five, similar in size to one of Harley Earl's divisional design staffs at General Motors. (Figures 10-22 through 10-26)[b] One reason that Gregorie could operate with such a relatively small staff was that everybody was allowed to do almost any task they wanted. He did not delineate responsibilities to the point where only clay modelers could do clay modeling or only designers could do designing. If a designer wanted to help with a clay model, or a modeler had an idea for a hubcap or a grille, they were welcome to contribute. "The innovation could come from anybody and not necessarily all from Gregorie," confirms Adams. "We could make drawings, templates, cut up letters for instrument panels, even paint the clay models. We could bring in things that we had done at home and throw them into the pot for consideration."

---

[b] Figures 10-22 through 10-26 are the only known photographs showing the entire Ford design department staff during Gregorie's tenure. The photographs were taken during the annual Christmas party for the department in 1945, one year before Gregorie's departure. Everyone is present except Ed Martin and E.T. Gregorie, both of whom were out of town at the time.

Figure 10-21. Leota Carroll was first woman designer in the Ford design department. She worked for Ross Cousins, who taught her how to make pastel renderings of automobiles. (Photo from the Collections of Henry Ford Museum & Greenfield Village)

Figure 10-22. Some of the Ford design department staff in 1945. Top row, left to right: Raymond Bliss, Arseno Iafrate. Third Row: Eric Ramstrum, Michael Homanick, William Fox, Herman Sommer, Charles Miller. Second row, left to right: Charles Stevenson, James Doyle, Harold Lucht, William Leverenz, Clyde Trombley, Ross McLean, Werner Framke, Arthur Karpeles, Cesare Testaguzza, George Tasman. Front row, left to right: Glen LaBreck, Gordon Fulton, Placid "Benny" Barbera, Richard "Dick" Beneicke. (Photo courtesy of John Najjar)

Figure 10-23. Some of the Ford design department staff in 1945. Top row, left to right: Robert Tarran, John Najjar, Harvey Baird, Robert Dolan, Elmer Gundy, Bruno Kolt. Second row, left to right: Ray Mecklenburg, George Wallis, Richard Skinner, Ben Kroll, Charles Miller, Lawrence Clark, Walter Keck, Leota Carroll, Ross Cousins, Clifford Summerville, Florence Peterson, Willys "Taillight Bill" Wagner. Front row, left to right: Frank Francis, Wayne Booth, James Morrison, Romeo D'Angelo. (Photo courtesy of John Najjar)

Figure 10-24. Some of the Ford design department staff in 1945. Top row, left to right: Hermann Brunn, Gerald England, James Huggins, T.C. Hobbs, Vern Breitmayer. Second row, left to right: F. McMaster, Francis Caspers, Thomas Hibbard, William Schmidt, Richard Baxter. Front row, left to right: Letha Allan, Elre Campbell, Beth O'Roke, Doris Dickason. (Photo courtesy of John Najjar)

Figure 10-25. Some of the Ford design department staff in 1945. Top row, left to right: Stephen Galla, Ted Regitko, Robert Todd, Emil Ganzenhuber, Fred Rosenau. Second row, left to right: Peter Ehlendt, Jay Haskins, Robert Doehler, John Cheek, Richard Velten, Victor Lang, Martin Regitko, John Dobben, Michael Bogre. Front row, left to right: James Levy, Rudie Welling, Bert Pugh. (Photo courtesy of John Najjar)

Figure 10-26. Some of the Ford design department staff in 1945. Top row, left to right: Walter Musial, Dutch Hoffman, Carrol Boice, Gordon Fulton, William Clark, Clyde Upton, Harry Preston, Robert Ferguson. Second row, left to right: John Duncan, Joseph Somogyi, Martin Noneman, Joseph McKee, John Rusk, John Hay, Anton Schuch, Anthony Wojtas. Front row, left to right: James Mearns, John McKenzie, James Lynch, Melvin Evans, Randall Osmon. (Photo courtesy of John Najjar)

With input from all, Gregorie's men came up with ideas that transformed the industry. They were first to integrate the radio speaker with the radio and put it on top of the instrument panel, they originated large plastic panels on the dash, they integrated the instrument cluster into the steering column, and they were first to conceal the running boards. As Gregorie says, "There were always fresh ideas from our design group."

In an era when dictatorial management was the rule, especially in the automotive industry and particularly at Ford Motor Company, Gregorie ran his department informally. "We were all a team," he proudly explains. He treated his men the way he wanted to be treated, and he related to them in the same way that a teacher relates to students: directing, encouraging, and critiquing. Gregorie kept the atmosphere light in the design department. "The whole thing was very simple," he says. "We just had a small staff, and Mr. Edsel Ford knew most of the boys in there by their first name—it was just a very friendly, folksy arrangement." The men called Gregorie by his boyhood nickname, Bob, and he called them by their nicknames. Gregorie called Emmett O'Rear, Emmett "O'Rump"; Willys Wagner, "Taillight Bill"; Placid Barbera, "Benny"; Eugene Adams, "Bud"; Tucker Madawick, "Tuck"; and, because of his German ancestry and thick accent, Dick Beneicke, "der Führer."

Francis Zawacki wanted to Americanize his name, as many Europeans did at that time, and began asking the others in the department to help him with suggestions. Smith, Miller, and Taylor all were proposed. However, Walter Kruke, Gregorie's good friend who liked to play court jester, began offering outlandish names such as Throckmorton or Rockelshouse. Kruke was kidding, but Zawacki took everything seriously.

"Francis would bring these lists of names to me to see if I'd approve any of them," chuckles Gregorie. "Finally, after a couple of weeks of this horseplay, I said, 'Hell, Francis, why don't you just call yourself Frank Francis? Make your last name Francis. Double Francis, see, double FF.' "

"That's what I'm going to do, Bob," replied Francis. "Thank you very much."

Another example of the informality within the design department was demonstrated when John Hay asked Gregorie if he could take off a couple extra days before Christmas to visit his parents in Ohio. Gregorie replied, "Well, it's up to you. It's your money!"

"We had an old draftsman in the department by the name of Rivard," relates Gregorie, "who kept an eye out for Henry Ford and the 'Russian council.' He wasn't very ambitious. He'd sit there at his desk...they'd all be coming back from lunch [the production men]—the Russian council—they'd stroll along up the aisle, and from about a thousand feet, you could see them coming. They'd walk along and be talking, and old Rivard would have an old blueprint, and he had a hole cut in it in the middle, and he could cock the blueprint so he could see through the hole without being noticed. I used to go up and tap him on the back and say, 'What does it look like today, Rivard?' "

Gregorie and Walter Kruke had been close friends since their early days with Crecelius. When they were single, they went out on the town together in Gregorie's magnificent 1930 LeBaron-bodied Stutz roadster or they went to football games in Ann Arbor in Gregorie's sporty 1920 Mercer Raceabout. (Figure 10-27) When the weather was nice, they went boating in Gregorie's beautiful black-hulled, 46-foot, double-masted ketch. When the boat needed a new diesel engine, Gregorie asked Kruke, along with Tucker Madawick and Ed Martin, to help him install it over a couple of weekends. "One day, a large wooden crate arrived at the receiving door of the design department," recalls Tucker Madawick. "Inside was a small four-cylinder diesel engine—no doubt, we thought, for our proposed new truck design. No, it turned out it was for Bob's sailing vessel, docked at his home on Grosse Ile. Bob paid us with cases of Budweiser beer to help him install it. The additional perk was the test trial runs up and down the Detroit River. What was the engine vibration like? How many knots could it turn out? Did it need a three- or four-bladed propeller? These

Figure 10-27. E.T. Gregorie poses next to his 1930 Stutz roadster outside Ann Arbor, Michigan, in 1934. When Gregorie sold this car, he kept the road lights and used them on the continental cars he designed for Edsel Ford. (Photo courtesy of E.T. Gregorie)

were challenging questions that Bob evaluated and answered. His landlubber design crew enjoyed just standing on the deck—sailing was never a prerequisite."

Edsel supported Gregorie's team approach. Edsel did not like formality and did not care to discuss anything with a group standing around him. He preferred talking alone with Gregorie in the privacy of Gregorie's sequestered office, deciding which way to go on a particular design. With talented artists such as Cousins on staff, Edsel still preferred to see one of Gregorie's quick sketches rather than fancy pastel renderings. "Edsel didn't care for flossy presentations," explains Gregorie. "He loved to be right there when I made a sketch—he liked to see it in the raw, so to speak. He was really charmed when I could just take an old blueprint and sketch up a hood ornament real quick-like—cartoon fashion. He loved that. He'd always accept something that was done quick like that." Because of this, Gregorie and his staff never gave any formal design presentations to Edsel, as was the practice at General Motors. All design decisions were made by Gregorie and Edsel, either in Gregorie's office or at the end of a drafting board in the design department.

Charley Sorensen could not get used to the democratic approach Gregorie was using in the design department. Although Gregorie now was part of the Ford executive team, Sorensen did not hesitate to throw an occasional barb at Gregorie to test his mettle. Emmett O'Rear vividly recalls a time in 1939 when Sorensen threw one of his famous temper tantrums against Gregorie.

"Sorensen walked into the design department about one o'clock," explains O'Rear, "and since Gregorie had not yet returned from lunch, Sorensen walked up to my drafting board, pulled up a stool, and watched me as I worked on a new grille design. About fifteen minutes later, Gregorie walked in, saw Sorensen, and came over. Gregorie said hello politely, but Sorensen ignored him for what seemed like hours, then finally asked, 'Is this going to be a stamped grille or a die casting?' Gregorie began to explain that by designing it one way, it could be a stamping; by designing it another way, it could be a die casting. Sorensen, not used to being lectured by anybody other than Henry Ford himself, became incensed. His face turned the color of liver, and he jumped up in disgust, knocking the stool over in the process. 'When are you going to get it through your goddamned head,' he yelled at Gregorie, 'that I don't give a shit about your ideas! All I want to know is, what does Edsel want?' Gregorie, not intimidated in the least, politely retorted, "I don't know what he wants. I haven't asked him yet.' 'Then you do it, and be quick about it!' demanded Sorensen as he spun around and walked out the door, slamming it so hard that I was surprised the glass didn't break. Gregorie just grinned, shrugged his shoulders, and walked back to his office. He wasn't shaken in the least."

Gregorie's personality would have forced him to stand his ground regardless of the environment, but it was easier for him to do so because he had Edsel's complete confidence and backing. Sorensen and the other Henry Ford men never accepted Gregorie as an equal. However, with Edsel fully supporting Gregorie, they had no choice but to work with Gregorie.

Although it took Gregorie a few years to staff his new design department, by late 1935 or early 1936 he and the small staff he had recruited had taken over all design activities for Ford Motor Company. During the next forty-eight months, Gregorie and his men redesigned the entire Ford line, created a new line of cars called the Mercury, developed the Lincoln Continental, and even designed a tractor at the bidding of old Henry! At the center of this flurry of creative activity was E.T. Gregorie. "He was *the* designer at Ford," explains John Najjar. "The rest of us did details." (Figure 10-28) Emmett O'Rear adds, "Gregorie spent fifty percent or more of his time at his desk, cranking out ideas for every aspect of the car from a styling point of view. He kept a ream (literally five hundred sheets) of eight-and-a-half by eleven white typing paper on the corner of his desk and grabbed off sheet after sheet, making these sketches. Shortly before lunch, he would come flying out with a handful of sketches and pass them out to Walt Kruke, John Walter, Bill Wagner, and Dick Beneicke.

Figure 10-28. This rare photograph of the Ford design department, taken in 1940, shows Gregorie, in the dark suit, leaning over the large drafting table, at work. Gregorie did the preponderance of design work himself and could be found frequently working on the boards with his other designers in this fashion. He also carried with him a clay modeling knife to make last-minute changes in a clay model if he desired. The designer in the foreground working on a quarter-size clay model is John Najjar. To the right is a sweep rack holding a variety of sweeps of various radii and a dummy of a U.S. male of average proportions, which was used in body development. The open, naturally lit area made an ideal work area for designing automobiles. (Photo courtesy of Herbert Gehr/LIFE Magazine © TIME, Inc.)

In my judgment, his ideas ran the gamut from bad to great, but the sheer volume was an unbelievable sight to behold. I've never in my lifetime seen a more fertile mind than Bob Gregorie's."

When he was not creating a design on paper, Gregorie would be studying an engineering drawing, arguing details with Sheldrick or Galamb, sculpting a scale model, or helping Beneicke work on a full-size model. "The clay modeling knife in Bob Gregorie's hand was a very artistic instrument," says O'Rear. Gregorie would also do design work at home, especially if a project was falling behind schedule. "I kept a pad at home, and I'd occasionally spend an evening sketching certain features that I had in my mind from a day at the office. I'd take a sketch pad and draw four or five ideas of something that was going on during the day, and take them in the next morning and give them to one or two of the boys and say, 'Pick up from here. See what you can do with it.'"

The one aspect of design that probably interested Henry more than Edsel was the commercial line of the company. Henry's pickups, trucks, and delivery vans all were highly functional; however, when Gregorie was finished with them, they were aesthetically pleasing as well. "We were always working on commercial stuff," says Gregorie, "but there was very little artistic effort put into it, if I can put it that way. It was usually pretty well controlled by a basic set of dimensions: the width of the body, wheel clearance, accommodating various size tires, and overhang, so we didn't have much leeway. But they always turned out passably decent-looking. Our efforts were primarily on surface detail. We usually made quarter-size models on the pickups, where we had the most amount of leeway and control."

The design process implemented at Ford by Gregorie was unique in the industry and enabled him to produce his landmark designs. The first step in his process was to develop an initial design concept, which Gregorie always did himself. Then a one-tenth or quarter-size clay model would be made. In the early years, Gregorie made these models himself, but as his apprentices became more experienced, he delegated much of this work to them. If the small-size clay model appeared promising, then a full-size clay model would be constructed. Edsel was involved in all these steps, but he usually waited until the full-size clay model began to take shape before providing his input. "In some instances," Gregorie recalls, "Edsel would take his hand and move it over the body design as though he was smearing the clay on the model himself to make his point." As soon as the full-size clay model was to Edsel's liking, Edsel would officially approve it, and the body engineers would take over from there.

After a clay model was completed but before the design was released to body engineering, Dick Beneicke and one or two design apprentices would take offsets at various points along the length of the clay model. Offsets are x, y, and z coordinates of almost every square inch of the formed clay surface. They are used to transform a three-dimensional design into a two-dimensional engineering drawing. At that time, these coordinates were measured with a modeling bridge, which is a large, sturdy structure that straddled the model. (Figure 10-29) Originally made of wood, these bridges were later precision-crafted from stainless steel at the Rouge plant. They had metal pointers protruding at right angles from the top beam and its side supports. The side pointer identified the y coordinate, and the top pointer identified the z coordinate. The x coordinate was simply the location of the bridge along the length of the model. Using the centerline of the front axle as their starting point, the clay modelers measured thousands of x, y, and z coordinates to describe a particular body design. The more curvaceous the panel, the more offset points were needed. For example, a highly contoured fender or hood would require that offsets be taken at every square inch, whereas all that was needed on a relatively flat roof panel was the centerline profile and offsets taken at ten-inch intervals. After all coordinates were taken and written in a book called a table of offsets, the body engineers could reproduce the entire surface of the car without ever having seen it.

The body engineering staff was not left in the dark concerning the designs Gregorie and his team were developing. In fact, the design department worked closely with body engineering. As a

Figure 10-29. Using a modeling bridge, Clyde Trombley takes precision measurements of the hood of one of Gregorie's "big Ford" designs. Original modeling bridges were made of wood, but all later bridges, such as this one, were made of stainless steel. From these modeling bridges, the designers would take x, y, and z coordinates and write them into a book called a table of offsets. From these coordinates, the body engineers would make two-dimensional engineering drawings. "These offsets were usually very accurate," explains Gregorie. "We couldn't be more than an eighth-inch off here or there. Once the lines were laid on the body draft full size, why, any irregularity would appear immediately, be easily checked out, and corrected." A large white sheet was draped over the models when they were not being worked on, to keep them out of the sight of curious eyes. In the background is a full-size clay model of a postwar delivery truck. (Photo courtesy of Ford Motor Company)

clay model neared completion, Gregorie would consult with Bill Pioch, the head tool and die man at the company, about the best place to put fender joints, hood joints, door hinge points, and other functional aspects of the body. "We wanted to clear any potential difficulty with the design before we handed it off to body engineering," explains Gregorie. "Most of my discussions with Bill were just more or less routine, but in many areas I wanted to get his approval before proceeding. It didn't take us long to arrive at a decision. If there were any little corrections needed, why, we didn't have to upset a thousand-man department to do it."

Gregorie had learned how to use modeling bridges and compile tables of offsets as a marine architect. "When I designed hulls of boats," he explains, "we took offsets at various places along the hull. Instead of having axle centerlines, fender contours, and roof shapes, we had water lines, deck lines, and sheer lines, and sections through the hull at various places. I just applied this process to the automobile business. I mean, designing automobile bodies was duck soup, compared to the outer body of an eighty-foot yacht!"

Prior to this sophisticated yet efficient process of layout development, explains Gregorie, "the old-timers used to make patterns, either out of cardboard or Masonite, to fit every ten-inch section of the body surface. They had to scrape, file, and finish those patterns until they precisely fit the body contour. Once all the patterns were made, the body engineers would transfer the forms of those templates to the full-size body draft."

At the next step in production, the head of the body drafting group, Clare Kramer, and his draftsmen became involved. They were located in one of the old Ford aircraft division hangars. Their responsibility, recalled Bud Adams, "was to put these sections on a full-size engineering draft, clean up the lines, and get the surfaces defined and trued up, so that they could take the draft back into the body engineering activity and do all the structure and the hinges and everything else that goes into a body." Tom Stephenson was one of those body engineers. "He had a couple of fingers missing from one hand," said Adams, "but, boy, he could lay these sections on, true the lines up, and come back and tell you that you had a little flat spot or a little hole in your model here or there, and, damn it, he was always right!" When the engineering drawings were completed, they were given to the pattern makers and then to the die makers.

This design process is what differentiated Gregorie's designs from those of Harley Earl. Instead of following Gregorie's design process of developing a design concept, starting a full-size clay model at the earliest convenience, and then working out the final shapes in clay, Harley Earl and his men took the opposite approach. They started with a full-size blackboard side view and perhaps end views or some cross sections. They would then complete the shape on the drafting board, using various "trade secret" templates to handle transitions in shape, to define the shapes between what they had established as basic sections. After completing these two steps, the design would be transferred from the two-dimensional drawing into a three-dimensional clay model. As John Najjar explains, "The designers at General Motors usually developed a good, flowing profile line from the rear fender to the rear door to the front door. But then that nice shape broke in plan view: the cowl and hood line took a sharp break inboard." Bud Adams concurs. "This type of approach," says Adams, "is what gave General Motors products their architectural hard feel, compared to Ford products. Our cars were generally more rounded and flowing and softer, not the stiff General Motors' type of appearance."

"When Martin Regitko first came in the door," continues Adams, "he had an armful of thick, old-fashioned French curves. They had sharp radii at the corners. They looked like the letter *L*. Evidently, these were what Regitko used regularly back East when he was doing custom-body design work—where the roof came over rather flat, and made a sharp turn, and came down the side of the body rather vertically. It was only a couple of days later that he had a vise attached to his drawing board, and at lunch time he was filing each one of these curves to suit the shapes that we were working with at the time. He reworked that whole family of curves until he had a new family of curves that suited the types of shapes that we were working on! We were pioneering with the more rounded roof and the angled tumblehome from there on down to the side of the body in generous curves." In essence, Regitko's modified French curves symbolized the evolution that Gregorie and Edsel were making in automobile design.

Eventually, each of Gregorie's designers had a complete set of French curves, or sweeps, with which to lay out designs, and these sweeps were modified as body designs became more streamlined and curvaceous. "We had a series of sweeps that ran four-hundred-inch and four-hundred-forty-inch radii," recalls Tucker Madawick. "If you were to do that full size, go out into the parking lot, get a line cord that's four hundred inches long, and swing an arc of it, that would give you a radius that we would have on the outside sheet metal of a Ford car, so when you leaned on it, it wouldn't 'oil can.'"

Many years had passed since Edsel Ford had first toyed with automotive design with his father's Model T runabout in 1911. Edsel would have preferred to have started a design department within the company much sooner than 1935. However, there is little doubt that Edsel had found in E.T. Gregorie the right man for the job. Edsel had considered Ray Dietrich years earlier but found Dietrich to be too arrogant and too much of a playboy for his taste. In retrospect, it is probably better that Edsel did not hire Dietrich because, after the custom-body business fell by the wayside, Dietrich never developed a suitable car for mass production; his forte was strictly custom bodies. Gregorie, with his combination of mechanical horse-sense and grasp of style, had a knack for the production car.

Edsel usually shied away from assertive people such as Gregorie, but he and Gregorie seemed to hit it off from the start. There were a number of reasons for this. First, both were young and had literally grown up with the automobile, and they understood the importance of styling to the customer. Second, both had similar tastes in automobile design. Third, Gregorie realized early that Edsel's desire was to build a design center of his own within the company, and he wanted to be a major player in it. Likewise,

Edsel knew that for the design department to have any chance of survival, he would need someone such as Gregorie, whose talent and temerity could stave off the interference received from the manufacturing men at Ford. Regardless of Edsel's innocuous intentions, Edsel knew that the mere existence of a design department would threaten men such as Pete Martin and Charley Sorensen, and even Henry Ford himself. They would consider the establishment of a design department as Edsel's attempt to usurp their power, rather than as a means of making the company more competitive. Although Edsel would have preferred a more reserved person with whom to start his design department, Gregorie's self-confidence and design talent fit Edsel's needs perfectly.

With their complementary needs and desires, Edsel and Gregorie quickly developed a mutual admiration for each other's personality and abilities. Concerning automobile design, they also were on equal terms. "Damn few people know how to design a car," exclaims Gregorie. "You have to get the heritage of an old Duesenberg, an old Mercedes." With such requirements, Gregorie could not have described his patron more accurately. Edsel knew automobiles and he knew design, and his new chief of design would not have to explain to him the difference between a Duesenberg and a Mercedes—or between a cabriolet and a roadster, for that matter. According to Ray Dietrich, when he first met Edsel, Dietrich was "thrown completely off guard by his extensive knowledge of the entire field of prestige cars, both European and American." When Dietrich asked Edsel questions about various aspects and qualities of different automobiles, Edsel always referred to "the different body styles by correct type of name according to the Society of Automotive Engineers [SAE] standard body nomenclature." Gregorie realized this fact also, and in his years of association with Edsel, he learned plenty about automobile styling and design from him.

Edsel and Gregorie had many similar tastes when it came to automobile design. Both men had a conservative bent, although Gregorie's was not as severe as Edsel's. They both appreciated the simple, straightforward line, and neither cared for excessive ornamentation on a car. Both men preferred the design of a car, as Gregorie says, to be "nice and trim and clean-looking." Above all, they both had a deep-rooted appreciation for the basic form. For them, the form of an automobile was paramount in design.

However, this is where their tastes diverged slightly. Gregorie preferred an "expressive, defined, and articulated" form. He liked the individual parts of an automobile body to flow together, while retaining their identities. On the other hand, Edsel liked forms with sharp contrasts and clear definition. Gregorie explains the difference this way. "Its similar to the difference between a sleek, delicate racing boat and a husky seagoing boat," says Gregorie. "Both have the ability to cut through the water, only one has a different purpose than the other."

Thus, their differences in design inclinations were slight, and any disagreements about which way a certain design should be developed were always resolved amicably between the two men. Edsel rarely gave a directive, preferring instead to solicit ideas and draw out new concepts through discussion. His unwillingness to sketch an idea himself forced the two to talk through a design. As they talked, Gregorie also sketched. As Gregorie sketched, Edsel made recommendations. "Can't we lighten that up a bit?" he would say. Or, "How about making this a little longer?" Or, "Would it do any harm to round that a little more?" His most servere criticism was, "Oh, that looks pukey, doesn't it?"

All these discussions between Edsel and Gregorie were candid and sincere because that is how Edsel liked to work. He hated formalities and meetings because he thought they wasted time and were too stifling. He preferred to work in a casual atmosphere, without pretense. "We worked like a country store," explains Gregorie. Edsel came into the design department almost every afternoon, and he was thoroughly aware of the designs on which all the men were working. If a decision was needed, Edsel would make it right there, Gregorie says, "just sitting at the end of a drafting board." Unless Gregorie had a specific objection to a particular way Edsel was directing him, he appreciated Edsel's design perspective and had enough respect for his artistic eye to follow Edsel's advice and "modify the design to a point where it pleased him." When he was satisfied with a design, Edsel would reconfirm the idea with Gregorie. "Does it look good to you?" he would ask. "Okay? All right, let's do it that way."

A project would usually be started by Edsel, but all design proposals and ideas would come from Gregorie and his staff. On his regular visits to the design department, Edsel would review the designers' ideas, make suggestions, and guide the men toward an overall design that was acceptable to him. That is probably why Edsel never offered a sketch of his own. If he did, it may have limited the designer's own talent, thus preventing the best possible design to unfold.

This process gave latitude to the Ford designers. Edsel believed that automobile design was an artistic endeavor, and he treated all his designers—especially Gregorie—as any beneficent patron would treat an artist. To Edsel, Gregorie was the artist, and Edsel merely gave Gregorie the means of creating. For his part, Gregorie appreciated Edsel's understanding of the design process and the support Gregorie and his designers received from Edsel. Such respect and understanding were "rare in the old automobile companies," Gregorie says.

If Gregorie could have changed one thing in his relationship with Edsel, it would have been Edsel's reserved nature. "Edsel didn't have an expressive personality," says Gregorie. "He was a very quiet, very reserved, unassuming man—always very solemn, very serious. He rarely exhibited enthusiasm, and he rarely exhibited distaste. It was very difficult to know, without very constant association with him, how his feelings ran. Maybe that was his privacy shield. I think he kept away from the masses that way. He didn't want to be influenced by people. He had his own way of determining whether a design or some aspect of a design met with his taste."

Gregorie also had his own ideas about how Ford design should evolve, but he was respectful and cautious as to how he presented his ideas to Edsel. "I had a feeling that he'd be offended if I became too pushy," says Gregorie. To help him convey his ideas as diplomatically as possible, Gregorie would sketch on a piece of vellum what he had in mind and then lay it over a design that he and Edsel were discussing to show Edsel what his design chief was thinking. That way, both of their ideas could be considered simultaneously. On many instances, Gregorie had to concede that Edsel's idea was superior to his. "But Edsel was an excellent critic," says Gregorie, "and he knew when something looked right. After all, a man doesn't have to be a designer to point out something and say, 'Look, the hood line doesn't look right,' or, 'I think the window is too wide.' He doesn't have to have the ability to take a pencil like I have."

The combination of Gregorie's artistic abilities and Edsel's support developed an uncanny bond, an almost ethereal relationship, between the great designer and his patron. "Even though there were two of us," explains Gregorie, "we thought as one. My hands became Edsel's tools in developing designs. I'm sure that sounds odd, but the designs that we developed over the years were *his* designs, not mine. It's difficult to put our working relationship into words, but in essence, I was able to put on paper and into clay the designs *he* was visualizing in *his* head. And I could do it without a lot of discussion between the two of us. Edsel never took a pencil in hand and drew an idea that he had—he depended on me to develop the design that he was visualizing. I sketched while he explained what he had in mind, and the design developed that way. Of course, we had similar tastes when it came to design, so if a design looked right to him, I had a feeling it looked right to me. And if he thought it looked right to me, unless there was something conspicuously wrong, why, it was all right with him. We never had any arguments over which way a design was to evolve. He'd make suggestions, and I'd make suggestions, and we'd alter the design until we were both satisfied with it. Throughout this process, however, it was as if he were directing my hands toward the designs he wanted. I know that sounds bizarre, but that's the best way I can explain the way Edsel and I worked together."

Edsel appreciated Gregorie's willingness to discuss design ideas with him and for not presenting arguments unnecessarily. Within a short time, Gregorie had become one of the precious few people within the Ford organization in whom Edsel could confide without fear of betrayal. The only other company employee with whom Edsel had had such a close relationship was his brother-in-law, Ernest Kanzler; however, Henry had Kanzler fired in 1926.

Because of Edsel's deep-rooted interest in design and his close relationship with Gregorie, the design department became Edsel's haven within the company. ("He came into the design department to deliberately avoid the old man," says Gregorie.) On numerous occasions, Edsel and Gregorie would climb into one of the sheet-covered prototype cars and talk. "It was cool in there," explains Gregorie, "and when Edsel didn't want to be disturbed, we'd sit in there on those nice, big sofa seats and talk about various things." Their discussions usually centered around design, but Edsel occasionally would bring in some magazine clippings of boats—in the same way as Edsel used to clip automobile pictures from magazines when he was a young man—and he and Gregorie would discuss their salient points. At other times, Edsel would confide in his chief designer about difficulties he was having in some other area of the company. However, Edsel never talked to Gregorie about the problems he was having with his father concerning design.

"It seemed to be sort of a relief to him to be able to talk to somebody without worrying about repercussions," explains Gregorie. "And I always felt very satisfied that he would take me into his confidence about those little incidents that took place within the company. But most of the time, Edsel would just like to go back in there, spend a half hour or an hour, and rest himself. He'd say, 'Gee, I wish I could sit here all afternoon.'"

Hiding in cloth-covered cars was the only way Edsel could avoid his father, because when Henry wanted to find his son, the old man would seek Edsel in the design department. Henry Ford's office was in the engineering lab, down the hall from Gregorie's office, whereas Edsel's office was four miles away in the Ford administration building on Schaefer Road. Edsel spent most of his mornings in his office doing paperwork, breaking at noon to meet with his father and other Ford executives for lunch in the executive dining room. (The dining room was located in a separate little building behind the engineering lab.) Hot meals were served there on a large circular table, and it wasn't long before these executive get-togethers became known as the "Round Table" meetings. (Figure 10-30)

Henry Ford could have held these meetings at the stately dining room atop the Rouge administration building where Edsel preferred, but that was too formal and businesslike for the old man. He preferred eating close to the experimental chassis and engine parts in the engineering lab and thus had this small dining room set up there.

Figure 10-30. The executive dining room was located in a separate building behind the northwest corner of the Ford engineering laboratory. Here, sitting at a round table, Ford executives would meet for lunch almost every day. Occasionally, Edsel or Henry would invite guests to have lunch with them, but E.T. Gregorie was never asked to join them. Shown from left to right are Edsel Ford, Henry Ford, Charles Sorensen, Pete Martin, and Al Wibel. (Photo from the Collections of Henry Ford Museum & Greenfield Village)

Edsel would drive to the engineering building and enter the little dining room through the front door, always arriving at precisely 12:15. By that time, all the other executives—Sorensen, Martin, Wibel, and others—were there. Henry was always the last to arrive, and he entered through the back door. Before being seated, Henry would circle the table counterclockwise and shake hands with everybody, ending with Edsel, who always sat to his father's right. Sorensen sat to Henry's left, and any guests that attended sat next to the person who brought them.

This round table protocol was started by Henry in lieu of formal committee meetings, which he abhorred. They were considered a replacement for those types of meetings, although business was rarely discussed. "Sometimes they would talk about business," recalled John Davis, Ford's sales manager and frequent guest, "but most of the time, it was largely a social group with just everyday chatter that one would hear at any formal luncheon table on subjects of the day." If business was mentioned, Sorensen usually started the discussions; even then, they were superficial. The "profound secrets" of the company, explained Henry Doss, a Ford branch manager, were not discussed there. Astonishingly, the weighty decisions concerning Ford Motor Company were made in private, impromptu meetings between Henry Ford and individual executives. For example, Henry might meet with Sorensen at the foundry or Martin at the Rouge plant. Because Henry rarely went to the administration building, he often met with Edsel to discuss important business matters wherever and whenever they ran into each other. Many employees recalled seeing Henry and Edsel take up *ad hoc* conversations in the engineering lab around a work table, in Greenfield Village on a park bench, or in Edsel's haven—the design department.

After lunch, the executives would disperse in their own directions. As for Edsel, he would walk down the long hall of the engineering lab to the other side of the building and visit with Gregorie. As the two discussed various projects on which the designers were working, old Henry would walk in, looking for Edsel. When Edsel saw Henry, he would excuse himself, walk over to his father, and go with him to one corner to talk—sometimes for hours. Edsel would sit on a drafting board, his hands under his legs and his feet dangling, and Henry would sit opposite him. "I always instructed the boys to stay clear, to find something else to do, and to give them all the privacy they wanted, when the old man came in," says Gregorie. "I was never bold enough to stay within earshot of their conversations, but I observed the pantomime between the two of them, and it never appeared pleasant. Neither one ever had a smile on his face." Emil Zoerlein, an engineer who worked closely for both Henry and Edsel for years, concurs with Gregorie's observations. He too had witnessed many discussions between father and son, and they never seemed cordial. "Edsel always had a worried expression...a painful expression," recalled Zoerlein. "It seems that Mr. [Henry] Ford was saying, 'Now Edsel, this is the way it's going to be.' "

The two Fords would meet in the design department, but they talked about everything except the design activities. "After their meetings," says Gregorie, "Edsel would briefly clue me in on the discussions that they had, and they never related to design."

Gregorie sadly recalls the interactions between Edsel and Henry Ford. "Old Henry Ford had absolutely no interest in the design and appearance of the car," he explains. "A lot of niceties that Edsel Ford wanted built into the product, Henry felt were superfluous, and that the design department was nothing more than a hotbed for problems. He never could understand why anybody would want to dress up in white smocks and build cars out of mud. And he tried awfully hard to stifle Edsel's involvement in design. For every new design that Edsel wanted to implement, the old man made him fight for it and the vast expense required. That's what Edsel suffered from ever since he pushed through the Model A. But, to Edsel's credit, he stuck to his guns, and pushed through all the designs that he and I developed. But Edsel paid dearly for it. After the meetings Edsel had with his father in my department, he'd come over and talk to me, and he was all shook up. It would take him a half hour or more to get his reasoning back. It was this constant wrangling with the old man that eventually wore Edsel down. It eventually killed him."

Except for John Tjaarda's Sterkenberg model that Briggs displayed at the Ford Exhibition of Progress in 1933 and again at the Chicago World's Fair in 1934, Edsel and Gregorie never became involved in producing show cars to test public reaction the way Harley Earl did at General Motors. However, Edsel allowed Gregorie to make one-off models—such as the continental models—or to revamp production models to test various design ideas. "Edsel always loved the idea of a specially built custom version of a Ford," says Gregorie. "The basic Ford, but a little different. Making these one-off cars reminded Edsel of the things he used to do to Model T's to make them sportier back in the old days.

He understood their usefulness." As with most things at Ford, no formal arrangement existed for these types of activities. The only unwritten requirement was that any design idea Gregorie wanted to pursue had to "contribute to the possibility of it being a production item or contribute to development work," recalls Gregorie.

Whenever Gregorie presented one of these ideas to Edsel, Edsel would look at it, "kind of wink, and say, 'Well, maybe we can arrange that. I'll see the powers that be. I'll see you next week.'" That was Edsel's way of telling Gregorie that he would think about it, but he rarely returned saying no. Gregorie once asked Edsel if he could put a landau top on a production Ford to see how it looked. "I told him what I had in mind," says Gregorie, "showed him a little sketch of it. He said, 'Okay, go ahead. We have to do those things to find out, don't we? Show it to me when you get through.' I took a '38 Ford Tudor, and I blanked out the rear quarter window. I sent it over to the Lincoln plant, and had it filled in and paneled with landau leather and all trimmed with pigskin, like on the custom-built Lincolns. I also had a special paint job put on it at the Lincoln plant. It was beautiful. It must have cost $50,000!" (Figure 10-31)

When the car was completed, Edsel drove it home one night and liked it very much. After that initial test drive, he gave the car to Gregorie, which became a ritual with many of these one-off designs. "It was one way Edsel could compensate me for the minuscule salary in those days," says Gregorie. Perhaps it was, but more likely it was Edsel's way of showing appreciation for the good work Gregorie was doing for him and the company. Edsel realized that Gregorie was, as John Najjar so succinctly puts it, "a designer's designer."

Figure 10-31. One of Gregorie's ideas to dress up the standard Ford was to blank out the top rear half of the body with leather, reminiscent of a custom-built Lincoln brougham. After viewing this mock-up, Edsel allowed Gregorie to make a prototype, but the idea went no further. (Photo from the Collections of Henry Ford Museum & Greenfield Village)

CHAPTER ELEVEN

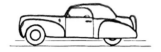

THE 1935-1937 FORDS
Years of Transition

For the 1935 model year, Larry Sheldrick made a subtle change to the Ford chassis that would influence Ford design for the next decade. His intent was admirable. To give the rear passengers a smoother and more comfortable ride, Sheldrick moved the rear seat forward off the rear axle kick-up. However, to retain the same amount of rear legroom, he moved the front seat forward, which in turn moved the cowl forward, and that forced the engine over the front axle. Unfortunately, Sheldrick kept the wheelbase the same, causing the cowl to be too close to the front axle. If Sheldrick had had any sense of styling, he would have moved the front axle forward the same amount he had moved the seats, cowl, and engine.

As Gregorie explains it, "These cars had a self-imposed deformity, which was the location of the front axle in relation to the bulk of the body—they didn't have enough wheelbase on the front end. The chassis engineers shoved the engine forward to where the spring was, forward of the front axle, which created a stubby effect in the front end and caused the illusion of the body trying to climb over the front axle. We had to utilize the space between the cowl and the front axle to get hood length, which was difficult to do. The only thing to relieve that was to pull the front wheels ahead six or eight inches."

If Sheldrick had increased the wheelbase by moving the front axle forward the same amount as he had moved the front and rear passenger space, every Ford from 1935 through 1948 could have had the same ideal proportions as the 1933 and 1934 Fords. Instead, every Ford produced during these years appeared nose-heavy.

"They had a heavy, bulbous, ahead-of-the-wheel appearance," says Gregorie. "There was a limitation as to how much overhang we could handle to minimize this effect. What we would like to have done was to have left the hood alone and pulled the front axle forward and added a little more length in the fender between the front door and the front wheel. The advantage of Sheldrick's change was that the cars now had more trunk space in the rear, and we were now able to get a very graceful slope on the rear end."

Gregorie noticed this affliction as soon as he saw the 1935 models, but he was too new in his position to approach either Sheldrick or Edsel about it. "I made quite a number of little sketches at that time—just doodling myself—which showed what a difference could be made by having a longer front end," he says. "Looking back on it now, I should have told Edsel that I wanted to experiment a little with it. But I didn't approach him with that idea. If I had tried to change things at that point, there would have been some hard words."

Besides, Gregorie was too busy working on continental cars for Edsel, and he never became involved with the design of the 1935 or 1936 Fords. These models were developed at Briggs Manufacturing under the direction of Ralph Roberts and his design group headed by John Tjaarda—with, of course, input from Edsel. (Figures 11-1 through 11-3) Phil Wright, a designer in Tjaarda's group, designed the stylish 1935 Ford line. Rhys Miller, whose first job as an automobile designer was working under Tjaarda in the spring of 1935, remembers seeing the clay model of the 1936 Ford almost finished when he joined Briggs. "The profile of the front end was done," recalls Miller, "but the grille design was still under development. John asked me to see what I could do with it. So I took a clay knife and carved out an oval-shaped grille which extended about halfway up the front and out into the catwalk area. It was a nice-looking design, but John felt that it wouldn't provide enough airflow to the radiator, which was tall and narrow." Holden Koto, another designer on Tjaarda's staff, eventually developed an appropriate facelift for the 1936 Ford. There is little doubt that Edsel also was involved. "You can see Edsel's influence in the '35 and '36 Fords," says Gregorie. "They have a Bentley look, and Edsel liked the design of the Bentley."

Gregorie is critical today when he describes the three Ford models with which he was not involved. "These '35 and '36s were spunky-looking, scrappy-looking. Pugnacious. They looked like

Figure 11-1. This quarter-size model of the design proposal for the 1935 Ford done at Briggs Manufacturing clearly shows the body, fender, and headlight design used in 1935 and 1936 Fords. However, the hood and grille are under development; the left half of the grille shows one idea, and the right half shows another. Renderings on the wall are comparing front-fender design with an ogee line (right) and no ogee line (left). An ogee line is a term used by designers for an S-shaped curve. The 1935 Fords had an ogee transition from the front fender to the running board, whereas the 1936 Fords did not. (Photo from the Collections of Henry Ford Museum & Greenfield Village)

Figure 11-2. This full-size wooden model of a 1935 Ford three-window coupe was designed and built by Briggs Manufacturing, inside the Briggs studio. "Before clay models," explains Placid "Benny" Barbera, "there were inside and outside wood models. They had seats, instrument panels, and everything— it was a complete car. And they didn't cost very much. In those days, they cost about $50,000. Briggs did an outstanding job building its wood models." (Photo from the Collections of Henry Ford Museum & Greenfield Village)

Figure 11-3. Shown here is a full-size wooden model of the final 1935 Ford design inside the design studio at Briggs Manufacturing. Phil Wright, a designer at Briggs, worked for Ralph Roberts and is credited for this handsome design. (Photo from the Collections of Henry Ford Museum & Greenfield Village)

a short guy with a pug nose." Gregorie uses the 1936 Ford three-window coupe as an illustration. (Figure 11-4) "Do you see how long the rear end appears in relation to the cabin or the passenger space? It looks like the body is trying to overtake the front end. Now, if you look at this same car from a three-quarter front view, it begins to look very decent proportionally. (Figure 11-5) In this view, the hood looks longer and the tail end looks shorter—foreshortens it a little bit."

The 1937 Ford is among the least revered Ford models, and Gregorie is quick to note that "the '37 was not done in our shop. That whole package was in process at Briggs at the time we were establishing the design department." (Figures 11-6 through 11-8) Always trying to save a penny here or a nickel there, Sorensen told Briggs to reduce the cost of Ford bodies. Briggs was already making bodies for Ford at rock-bottom prices, and when they were ordered to lower prices even more,

Figure 11-4. This profile of the 1936 Ford coupe clearly shows the stubby front end that Gregorie abhorred, but about which he could do nothing. However, Gregorie did a marvelous job in subsequent years to minimize this effect. Comparing this car to Gregorie's 1939 Ford coupe design (Chapter 14, Figure 14-1), Gregorie achieved a better-proportioned automobile, although the chassis was the same. "We were trying to get a longer flow line in profile in these cars," explains Gregorie. (Photo from the Collections of Henry Ford Museum & Greenfield Village)

Figure 11-5. In this three-quarter front view of the 1936 Ford coupe, the stubby front-end effect is minimized, making the body appear better proportioned. (Photo from the Collections of Henry Ford Museum & Greenfield Village)

Another shortcoming of the 1937 Ford was that the front fenders had an extremely long draw from the outside edge of the fender to the inside edge, causing a high parts-rejection rate. This caused undue grief for manufacturing and the consternation of Charley Sorensen. Sorensen talked to Edsel about changing the design so the fenders could be more easily manufactured. Edsel was receptive, but he told Sorensen that nothing could be designed and ready for production until at least the 1939 model year.

The only saving grace of the 1937 model was the front end, and that was borrowed from Gregorie's 1936 Zephyr. The 1937 Ford was designed in late 1935 and early 1936—around the time the Zephyr was introduced—and Edsel liked the Zephyr's front end so much that he ordered Briggs to incorporate it into the 1937 Ford.

the only thing they could think of was to make the bodies smaller. In other words, they would use less sheet metal. The result was a styling disaster. The front end was already pugnacious, and now the rear end was shortened, too. This gave the 1937 models a "pinched" appearance. "They looked hungry, a little bony," describes Gregorie. "Everything looked skimpy: the body, the bumper sections, even the hardware."

The 1937 Ford was the last Ford model designed with the help of an outside body supplier. (Figure 11-9) In the summer of 1935, Gregorie (who was only twenty-seven years old) and his group of fledgling designers began work on the all-new 1938 Fords. Edsel finally realized his dream of having all design work done within Ford Motor Company, and Gregorie dug in for the duration.

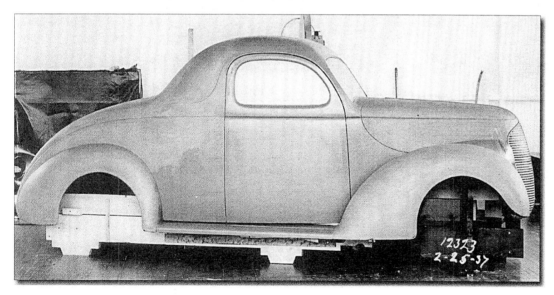

Figure 11-6. This was a design proposal for the 1937 Ford coupe by Briggs. Note how the rear of the front fender extends into the door, a design trick conceived by the Briggs designers to develop the illusion of more length in the front end. Edsel did not like this idea. The body design is similar to the Briggs 1935 and 1936 designs and would probably have gone into production in this form were it not for Sorensen's demands to decrease the cost of the bodies. When that occurred, the entire design was revamped. (The dates on the photographs are not the dates on which the clay model was done, which was some months earlier; rather, they are the dates on which the photographs were taken.) (Photos from the Collections of Henry Ford Museum & Greenfield Village)

Figure 11-7. This is an early clay model of the 1937 Ford done at Briggs Manufacturing. At Edsel's insistence, the Briggs designers incorporated Gregorie's 1936 Lincoln-Zephyr grille design into this Ford design. (Photo from the Collections of Henry Ford Museum & Greenfield Village)

The 1935–1937 Fords

Figure 11-8. This is the final version of the Briggs 1937 Ford design. Briggs had finally begun to use clay in building its models instead of wood. Briggs' expertise came from Dick Beneicke. (Photos from the Collections of Henry Ford Museum & Greenfield Village)

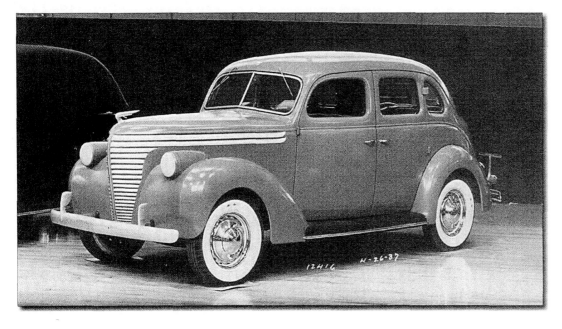

Figure 11-9. These are full-size design proposals for 1937 or 1938 Fords. Based on their appearance, these cars were probably developed by Amos Northup at Murray Corporation. Since Murray had become involved with the Model A ten years earlier, it tried in vain to obtain additional design contracts from Ford. None of these cars was ever produced. (Photos from the Collections of Henry Ford Museum & Greenfield Village)

CHAPTER TWELVE

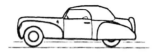

THE 1938 LINCOLN-ZEPHYR
"It's Going to Ruin Us!"

Soon after Gregorie and his staff began work on the 1938 Fords, the company started receiving complaints from dealers and customers in the Southwest, claiming that the new Zephyr overheated too easily. The condition was caused by a combination of mechanical design and styling. Because of Gregorie's front-end design, the engineers were forced to install a tall, narrow radiator behind the vertical, V-shaped, pointy grille. This arrangement did not allow the engine fan, which was mounted low, right off the engine crank, to pull enough air across the entire surface area of the radiator to cool the engine. Frank Johnson, the chief engineer at Lincoln, asked Gregorie if something could be done to the front end of the car to allow more air to pass through the radiator. (It probably would have been as easy to relocate the engine fan, but Henry liked its simplicity—no belts and no pulleys—so that was out of the question.) Gregorie had never become comfortable with the vertical front end he had designed for the Zephyr anyway, so he was more than happy to see what he could do to help.

After lunch one afternoon, Gregorie walked up to a Zephyr chassis in the engineering lab and started taking its measurements, and he determined that the radiator would fit horizontally between the frame rails. Mounting the radiator on its side would allow the fan to pull more air across more surface area of the radiator, thus providing more cooling. "So I had the boys change the radiator around and whip together some rough sheet-metal work around the radiator and just piece it together with little pop rivets," explains Gregorie. "It was just a rough bunch of sheet tin. But the opening we cut in it to allow air to flow through was low and crosswise in front of the core. We put it in the wind tunnel, and we ran it, and, gee, it cooled fine." When Gregorie showed Edsel his idea, Edsel said, "Well, let's go ahead and see what we can do with it." So Gregorie had Dick Beneicke construct a front end "with two grilles, quite low, and it turned out to be right nice-looking. Of course, Edsel okayed it right away for the '38 Zephyr." (Figures 12-1 through 12-4)

Figure 12-1. Gregorie's horizontal grille treatment on the 1938 Lincoln-Zephyr was the first of its kind and a styling sensation that caught the industry by surprise. General Motors and Packard soon followed suit. (Photo from the Collections of Henry Ford Museum & Greenfield Village)

Figure 12-2. Here is another concept for the 1938 Zephyr grille. "That center piece of the grille, well, there isn't enough of it to be a grille really," says Gregorie. "It's a little five- or six-inch piece that tapered down to nothing, not that we couldn't have used the extra cooling!" (Photo from the Collections of Henry Ford Museum & Greenfield Village)

Figure 12-3. This was a design proposal for the 1938 Lincoln-Zephyr. "This is when we were changing the appearance of the '36–'37 Zephyr," explains Gregorie. (Photo from the Collections of Henry Ford Museum & Greenfield Village)

The original design of the Zephyr was beautiful in its own right, but Gregorie's new design was so stunning that it became an overnight styling sensation. Its two-piece grille had fine, horizontal bars, the first of its kind and the envy of the industry. Packard copied it immediately. When Harley Earl, Gregorie's rival at General Motors, was shown a photograph of the new front end of the Zephyr, Earl became enraged. "How'd we miss on that one?" Earl remonstrated. "It's going to ruin us!" In response to Gregorie's styling coup, Earl ordered horizontal grille designs for all General Motors cars as quickly as possible. (Earl's first design with a horizontal grille was on the 1939 Buick.) "All the other manufacturers thought it was a styling deal," explains Gregorie. "They didn't know I did it to correct this damn cooling problem!" In fact, Gregorie reverted to a more vertical grille treatment on the 1939 Zephyr. He retained the basic outer shape of the two-piece grille of the 1938 model, but he replaced its fine horizontal bars with fine vertical bars, giving the grille a more flowing effect, as a yacht cutting through water.

Figure 12-4. This mock-up was developed to determine the most appropriate grille shapes for the 1938 Zephyr. "There was never a center grille as shown," explains Gregorie. "With this model, it is apparent how much better the vertical grille bars look than the horizontal grille bars. This was a trial-and-error thing. It shows some of the antagonizing things we went through in developing this design." (Photo from the Collections of Henry Ford Museum & Greenfield Village)

CHAPTER THIRTEEN

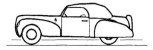

THE 1939 LINCOLN-ZEPHYR CONTINENTAL
"The Most Beautiful Car in the World"

Edsel may have put the sports car idea on the back burner, but Gregorie had not forgotten about Edsel's desire for a continental-type car. Throughout the whole period of Zephyr development, hardly a day passed when some idea for such an automobile did not cross Gregorie's mind. However, Gregorie also was preoccupied with a more pressing necessity. The Model K Lincoln—the "big K Lincoln," as Gregorie called it—had been in production almost unchanged since early 1931 and was extremely outdated. This was especially important because the Model K was the most prestigious car offered by the Ford Motor Company. Add to that the fact that the Zephyr was not setting any sales records for Lincoln, and Gregorie was understandably concerned that Lincoln might be on its last legs. From Gregorie's perspective—and from Edsel's as well—Lincoln needed a "spiritual boost" to lift it from its doldrums. What Lincoln needed, thought Gregorie, was a classy car that surpassed even the streamlined Zephyr.

As Gregorie was thinking about designs for both Edsel's continental car and a replacement for the Model K, a concept popped into his head one evening in the fall of 1938. Because he and Edsel could not add a new chassis to the Ford production mix, Gregorie wondered whether one of the chassis currently being manufactured by Lincoln could be utilized in a new sporty vehicle. The answer became readily apparent. The Zephyr had a unique chassis unto itself, with a shallow floor pan and low curb height. Perhaps a new car could be designed around it that was sporty, or elegant, or both. If so, then the Zephyr's suspension, engine, steering gear, and other mechanical parts probably could be used as well. Gregorie thought about it most of the night, and when he arrived at his office the next morning, he sketched his idea.

Gregorie asked his right-hand man, Ed Martin, to pull out some one-tenth-scale drawings of a Lincoln-Zephyr four-door sedan. "As I sat beside him," recalled Martin, "he [Gregorie] took out a piece of vellum, put it over the sketch of the Lincoln-Zephyr, and began drawing. After several minutes of silence, during which he kept drawing, he told me that since it was impossible to build a sports car on a Ford chassis, he would suggest to Mr. Edsel Ford that we build a special sports car on a Lincoln-Zephyr chassis." (Figure 13-1)

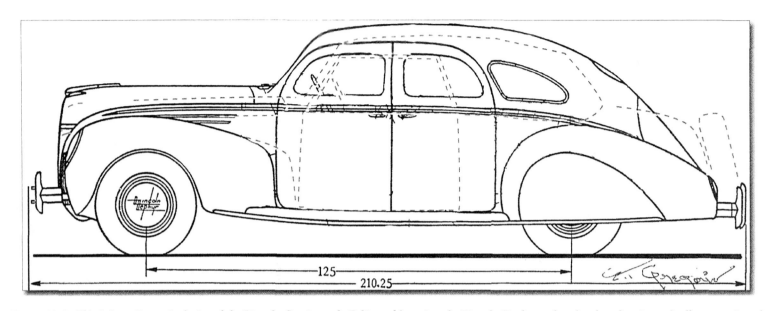

Figure 13-1. This is how Gregorie designed the Lincoln Continental. Taking a blueprint of a Lincoln-Zephyr sedan, he placed a piece of vellum over it and sketched a profile that would become the Continental. Gregorie completed it in only thirty minutes. His new design appeared longer and lower, although it had the same wheelbase as the Zephyr sedan. "The profile was pleasing," admits Gregorie, "and that crude little sketch that I made of it is what sold Edsel on the concept. He loved those simple sketches." (Photo courtesy of Warren Williams)

Although the Zephyr was a comparatively long and low car, the new design that Gregorie sketched appeared even longer and lower. In reality, it was not any longer or lower to the ground than the Zephyr because the same chassis was being utilized. However, after Gregorie had moved the windshield back, lowered the roof line, lengthened the fenders, and made the hood even longer and lower, the vehicle appeared significantly longer and more sleek than the streamlined Zephyr. In less than an hour, Gregorie had roughed out the design that eventually became the Continental. "When I tell people that I designed the Continental in thirty-five minutes or so, why, that's the honest truth!" Gregorie exclaims proudly.

That afternoon, Edsel dropped by the design department as he normally did after lunch. As he approached his chief designer's office, Gregorie pulled the crude sketch off his drawing board and held it up for Edsel to see. "How do you like this?" Gregorie asked.

"Oh, boy!" exclaimed Edsel. "That looks great! That's it! Don't change a line on it! How fast can you have one built?"

Because Edsel's Florida vacation was planned in March, that became the target date for completion. To ensure that none of the lines of the design were changed, Gregorie had Bud Adams immediately take his sketch and glue it to a piece of Masonite, cut it out on a coping saw, and set it up in a small modeling bridge. Then, in his office and with help from Adams, Gregorie began carving a one-tenth-size three-dimensional clay model. (Figure 13-2) As their work progressed, "Edsel would periodically come in very quietly to review the model," recalled Adams. "If Mr. Gregorie wasn't there, I would scurry out and get him. I would then sidle up behind the two as close as possible so I could hear their comments firsthand." The model was completed in approximately a week, and after having it painted Edsel's favorite gunmetal gray, Gregorie realized that he "had a real nice-looking car." So did Edsel, and the prototype was begun.

Before the craftsmen could start work, engineering drawings had to be made. Bud Adams transferred the profile of the clay model onto a one-tenth-scale drawing (Figure 13-3), and Martin Regitko, Gregorie's top body draftsman, used both the small clay model and Adams's scale drawing to develop a full-size engineering drawing that the model shop used to construct hammer forms for the fenders and body panels. "Other details of construction were considered and worked out to meet any objections that might arise," recalled Martin. "Men from the Lincoln plant—Harold Robinson, Walter Street, and Mr. Bartholomew—were consulted." At that point, the prototype car was built.

This was a highly unusual procedure for building a prototype. Typically, a watercolor rendering of an ideation was made to get a feel of the design proposal. If that appeared interesting, additional sketches might be made to provide additional perspectives of the

Figure 13-2. As this photograph of the clay model of the Continental attests, the simple profile that caught Edsel Ford's eye became elegant in three-dimensional form. "Once we had the clay model," describes Gregorie, "why, we all realized that we had a real nice-looking car!" The spare tire shown on this model was not part of the original model—it was added after the prototype was completed to bring the model up-to-date. (Photo from the Collections of Henry Ford Museum & Greenfield Village)

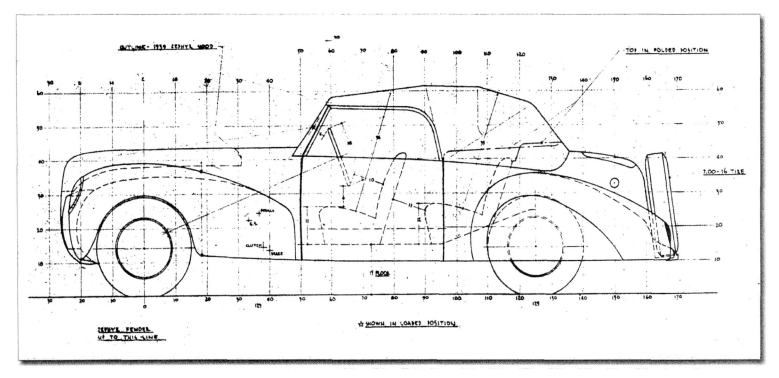

Figure 13-3. Bud Adams, one of the Trade School "boys," developed this scale drawing of the Continental by taking points off the one-tenth-scale model that he and Gregorie had made earlier. From this drawing, Martin Regitko developed the full-size engineering drawing that the body engineers used to hand-build the prototype. (Photo from the Collections of Henry Ford Museum & Greenfield Village)

design. Then a full-size clay model would be built to see the design in full-blown, three-dimensional size. (Figures 13-4 and 13-5) All these steps would have taken at least three months to complete, *if* everything fell into place. Even at this late stage in the process, the design could be significantly altered or even scrapped. However, the Continental bypassed all those steps. Watercolor renderings, presentation sketches, and a full-size clay model were not made. Only the rudimentary profile sketch, the small clay model, and the abbreviated engineering drawing were made before the prototype was built. A full-size clay model and all necessary engineering drawings were not made until after Edsel had decided to put the car into production.

To meet Edsel's deadline, Gregorie took a production 1939 Zephyr sedan and had it revamped by hand into a new automobile, based on his fresh design. The beautiful front end of the Zephyr, which Gregorie had redesigned for the 1939 model year, was left intact; however, many other body parts were either revamped or discarded. The cowl was retained, but four inches were sectioned from it and the doors to give the body a lower

Figure 13-4. This full-size clay model of the 1940 Lincoln Continental was made after the prototype was built, a sequence opposite the normal design process. By using shortcut methods, Gregorie and Edsel placed the Continental into production less than a year after Gregorie conceived its design. (Photo from the Collections of Henry Ford Museum & Greenfield Village)

Figure 13-5. On this early clay model of the 1939 Lincoln-Zephyr, note the different detail of the grille bars on the left side and on the right side. The design on the driver's side of the car was chosen for production on the Zephyr, as well as on the Continental. (Photo from the Collections of Henry Ford Museum & Greenfield Village)

The form was already precast. We happened to have a nice-looking front end, and the fenders were nice, and by the time we squished it down and raked the windshield a little more—why, you had your car! It just happened. It was a happy combination. When you stop to think of the agonizing that goes on when you attempt to do that from scratch with airbrush drawings and models and all that type of thing, I think we came out with something that was damn striking without all those shenanigans. It showed what could be done. Look at it this way: it was an unusual way to do things, and we at Ford at that time frequently did things in an unusual way, and we frequently came out with some unusual cars. I don't think we would have gotten a car as attractive as the Continental turned out if we had hashed around with all sorts of meetings and conferences and all that other 'blah, blah.' Edsel liked that. He had a sense of liking things done that way. In other words, the Continental could only have happened at Ford."

Gregorie designed the Continental in the design department, but the prototype was built at the Lincoln assembly plant ten miles away. Gregorie frequently would go to the plant to monitor the construction, and Edsel also would digress from his normal routine to go to the plant several times a week to follow the progress. During these visits, two of the most striking and influential design features of the car were developed.

profile. The hood and all the fenders were replaced with hand-fabricated ones, which were much longer than those of the standard Zephyr.

"There weren't months and months of drawing and building models," says Gregorie. "I put together so many cars like that, just out of odds and ends, so it became just a backyard project.

When Gregorie had made the original sketch, he drew an integral rear trunk, giving little thought to its functionality. However, by the time the prototype was almost finished, Gregorie realized the trunk was too small to hold a spare tire. The deadline was rapidly approaching, and Gregorie dreaded the possibility of having to revamp the trunk area at this late date.

Gregorie cornered Edsel one afternoon and explained his dilemma. The two men talked for what seemed to be hours, but they finally arrived at a solution. What appeared to be a classic decision of placing function over form was actually a compromise to keep the project on schedule. Instead of redoing the trunk area to hold a spare tire, Edsel and Gregorie decided to mount the spare tire on the outside of the trunk lid. They had no other alternative; after all, it was the only place available on the car on which to put the tire. "It surely wasn't a styling twist," Gregorie says with a smile.

The other design feature that also became a Continental trademark resulted from expediency instead of deliberate design intent. Edsel did not care for the use of much chrome, says Gregorie, such as "big, grotesque hood ornaments or great big, thick bumpers" which were becoming prevalent during this time. In maintaining this philosophy, Edsel instructed Gregorie to avoid adding any kind of trim to the body—to omit even the simplest chrome piece.

Gregorie did not question Edsel's request because it made his task much easier. Therefore, he had the car built completely bereft of any kind of body trim, keeping it "clean and neat." This allowed the fine surface lines of the basic design to remain uncluttered by strips of chrome or stainless steel. The few chromed pieces on the car were the hood ornament, the door handles, and the trunk hinges. Gregorie's illustrator, Ross Cousins, designed the beautiful hood ornament in an Art Deco style. Cousins emphasizes that "1939 was the year of the World's Fair, and I got my inspiration for the sphere and lance design from brochures I had seen of the exhibition."

After the Zephyr body was disassembled, cut, and sectioned, and the new body parts were added, all imperfections, rough surfaces, and transitional areas were filled in with lead and smoothed. "I shudder to think how much the car weighed," explained Edsel's youngest son, Bill, who drove the prototype a few times. "I don't know how much lead was in the thing, but it was really a lead sled. It took about a mile to stop the car after you got it up to sixty-five. The brakes were totally inadequate for its weight."

Because the car was not built in the design department, the other Ford designers were completely oblivious to the progress of the project. The first time any of them saw the finished vehicle was when the prototype was finally completed and moved from the Lincoln plant to the design department. (Figures 13-6 through 13-8) "We watched with amazement," recalls Bob Thomas, a junior designer at the time, "as the car was moved into the studio. It was a puzzlement. What was the squared-off rear end with an add-on spare tire? That hood—it was a foot longer than the Zephyr's! Has Gregorie lost his mind?" But after overcoming his initial shock and analyzing the design more closely, Thomas and the other Ford designers realized that Gregorie had made "a beautifully proportioned car."

The car was shipped from Detroit to Florida in time for Edsel's vacation, where he drove it around Palm Beach to the delight of onlookers. The car was the hit of the winter season among the Gold Coast *beau monde*. Edsel received such acclaim for the car that he telephoned Gregorie two weeks later, full of excitement. "Gosh!" he exclaimed, "I've driven this car around Palm Beach, and I could sell a thousand of them down here right away. You'd better start on another prototype right away for engineering and test purposes, and get over to the Lincoln plant and see what you can do to set up arrangements for limited production."

At that moment, Edsel was thinking of producing the new car in a limited production run of only one or two thousand units, something he used to do with custom-bodied Lincolns over a decade earlier. However, between the time of his phone call and

Figure 13-6. On a cold February day in 1939, Edsel Ford asked George Ebling, the Ford Motor Company resident photographer, to go to the Lincoln plant and take some pictures of the car Gregorie called the Lincoln-Zephyr special convertible coupe. Ebling backed the car out of the Lincoln factory where it was built, out into the barren landscape of Detroit in winter. Juxtaposed against an empty lot with freight cars in the background, Ebling began to record for posterity the innocuous beginnings of what was to become "the most beautiful car in the world." (Photo from the Collections of Henry Ford Museum & Greenfield Village)

Figure 13-7. Shown here is a left side profile of Gregorie's fabulous Lincoln Continental. Gregorie says that the profile of this car "was just about perfect." (Photo courtesy of Ford Motor Company)

Figure 13-8. This photograph shows the simple but elegant front end of the soon-to-be Lincoln Continental. Instead of having the prowling vertical grille as shown, the car narrowly escaped having a horizontal grille design. The light, delicate-looking bumpers were Edsel's preference, as was the lack of decorative trim. (Photo courtesy of Ford Motor Company)

the time he returned to Detroit a month later, Edsel had decided to incorporate the car as part of regular production at the Lincoln plant, commencing with the 1940 model year.

Initially, Gregorie referred to the new car simply as the "special convertible coupe." However, before the car went into production, Edsel decided to make it the top of the Lincoln-Zephyr line and christened the car the Continental to depict its dignified but fast, sporty, and exhilarating design. Formally speaking, the name of the new car was the Lincoln-Zephyr Continental. (The hub caps on the prototype had Lincoln-Zephyr stamped into them.) The car did not lose the Lincoln-Zephyr moniker until 1941, when it became its own distinct series in the Lincoln line.

Only 404 Continentals—350 cabriolets and 54 coupes—were built during the 1940 model year, and all of those were essentially hand-built. (Figure 13-9) That volume was much too low to justify spending the hundreds of thousands of dollars it would have cost to tool up for the new model. Thus, the Lincoln plant relied on its hundreds of craftsmen and artisans left over from the custom-body days to build the limited number of Continentals. As in the old days of building custom bodies, they used wood forms and Artz presses to form the hood, fenders, and doors instead of costlier steel dies. They also substituted aluminum castings where steel stampings would normally be used. In areas that needed additional finessing or a smoother finish, lead was used to fill any imperfections. Production tooling was not used to make the luxurious model until the 1946 Continental went into production after World War II.

Production of the Continental was low, but it served an important purpose in the Ford lineup. From Gregorie's point of view, the Continental saved the Lincoln line. Sales of the Model K Lincoln had dropped dramatically—from a high of approximately 3,500 units in 1931 to only a little more than 200 units in 1939. The Model K Lincoln was dropped that year. (Figure 13-10)

"If it hadn't been for the Continental," says Gregorie, "the Lincoln name would have fallen into a precarious position after the exclusive K Model was phased out of the picture. With the K Model gone, there was nothing to carry the Lincoln name with

Figure 13-9. Here, designer John Najjar works on a model of Henry Ford's so-called soy bean car, a vehicle in which the body was composed of plastic made from soy beans. In the background is a full-size clay model of the Lincoln Continental coupe. (Photo courtesy of John Najjar)

any sort of prestige. The Lincoln-Zephyr never had prestige—it was a twelve-hundred-dollar competitor to Buick and Oldsmobile. So when the Continental came along, it gave Edsel Ford a good prestige boost with the type of car that he was primarily interested in, and it became an exclusive carry-over for the Lincoln name, which carried it right up to the war. Ironically, Edsel's interest in adding a sports car to the Ford line was to enhance the Ford name. Nobody ever thought of making it a Lincoln until we didn't have a Lincoln!"

Figure 13-10. Shown here is a design exercise for the Model K Lincoln. "This was Edsel's attempt at resurrecting the big K Model," explains Gregorie. "The front end looks very suitable for a truck—I did truck front ends very similar to this. Mechanically, the Model K was hopelessly out of date, so we never got any further than a full-size clay model. We didn't fuss with it very long." The Lincoln Continental replaced the Model K Lincoln in 1940. (Photos from the Collections of Henry Ford Museum & Greenfield Village)

Figure 13-10. (continued)

Did Gregorie's Continental typify the continental car that Edsel was seeking? That will never be known, for Edsel never solidified what he meant by "continental car." Edsel died before having a chance to pursue a smaller, sportier car, if that is what he ultimately had in mind. All Edsel ever relayed to Gregorie in specific terms was that he wanted a car that was "long, low, and rakish," and the Continental obviously fulfilled those requirements. However, so did all of Gregorie's continental cars, and so did the English Jensens. The fact that Edsel was willing to produce a car the size of Gregorie's sporty 1934 continental car leads us to believe that Edsel was more inclined toward a Ford-sized sports car than a Zephyr-sized one.

Thus, the eventual creation of the fabulous Continental resulted from a serendipitous chain of events. The car was not created with any specific guidelines, but it vaguely emerged from Edsel's desire for a sport car for the Ford line and from Gregorie wanting to jazz up the Lincoln line. In its own way, the Continental fulfilled both men's desires.

Today, Gregorie has mixed emotions about the Continental. "Its profile, I feel, was just about perfect," Gregorie says. "It was just so simple, it couldn't help but look nice. We happened to have a nice-looking front end, and the fenders were nice, but I never prized it as a beautiful piece of work. I thought it was kind of weak in the rear end. I didn't care too much for the pinched-in rear end. It reminded me a little bit of a dog with his tail between his legs. I always thought it looked a little hungry. And the bumpers were a little too delicate. I would have liked a little more beef in them. But, of course, Edsel liked those delicate bumpers."

Gregorie is surprised about all the acclaim he has received over this one car. "When you think of the trucks and buses and the thousands of other vehicles we designed," says Gregorie, "I don't understand why people single out the Continental. But it was spectacular, and it was more of a deviation from conventional design—let's put it that way—for anything that ever went into production. As it developed, it became very popular, and I guess it'll go down as one of my better-known activities."

That is an understatement. Gregorie's Continental has been the only American luxury car ever honored for design excellence by the Museum of Modern Art in New York. Likewise, the brilliant architect Frank Lloyd Wright told a reporter in 1940 that he thought Gregorie's Continental was "the most beautiful car in the world." Few automobile designers can claim such distinction.

CHAPTER FOURTEEN

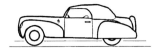

THE 1938-1940 FORDS

"From an Old Car to a New Car"

During the last half of 1935 and into 1936, Gregorie and his new design staff completely revamped the entire Ford line. Introduced as the 1938 models in the fall of 1937, these were the first production Fords with which Gregorie was involved since the successful 1933 and 1934 models.

For the first time, Gregorie was able to delegate design activities to a staff. Dick Beneicke and Bud Adams worked on the body surfaces. Placid "Benny" Barbera designed the taillights, license plate brackets, and trunk handles, working with "Taillight" Bill Wagner, who did the design sketches of these items. The profiles of these cars were done by Gregorie's good friend, Ed Martin. "Ed spent almost one hundred percent of his time working on the blackboard," explains Emmett O'Rear, "and that's where these graceful profiles were originated." (Figures 14-1 and 14-2) However, the keystones of these handsome designs were their front ends. Bruno Kolt made the perspective sketches of the front-end sheet metal, and O'Rear made the clay representations. The ideas came from Gregorie.

The 1938 model year was also the first time Ford offered two distinct series in the Ford line: a Standard and a Deluxe. (Gregorie, in his New York accent, says "Dee-looks.") Ford had been offering Standard and Deluxe models since 1930, but until now, the only difference between the two models was the addition of some stainless steel trim on the Deluxe model. Beginning with the 1938 models, the distinctions between the Standard and the Deluxe became dramatic. Edsel wanted more than added trim to distinguish a Standard model from a Deluxe—he wanted cars with completely distinct appearances.

Gaining more distinction between the Standard and the Deluxe was another attempt by Edsel to offer the public a wide array of cars at various price levels, similar to what General Motors was doing. With this new marketing strategy, the first-time Ford buyer could now buy the less-expensive Ford Standard, while the current Ford owner could move up to the more expensive and more attractive Ford Deluxe or even the streamlined Zephyr. Of course, the top Ford offering was the Lincoln. What was missing from

Figure 14-1. This stunning profile of the 1939 Ford coupe shows one of Gregorie's most handsome designs. The passenger compartment relative to the rear axle is what enabled Gregorie to develop the flowing rear end. However, from Gregorie's viewpoint, the front end was too foreshortened. (Photo from the Collections of Henry Ford Museum & Greenfield Village)

Figure 14-2. This profile shows the handsome 1940 Ford Deluxe convertible. "I like a sharp effect in the convertible top—put just enough padding in the top to make sure it doesn't look like a hungry cow," explains Gregorie. "Regitko liked more padding. I'd tell him, 'It looks like you got a mattress under there!'" (Photo from the Collections of Henry Ford Museum & Greenfield Village)

model, and he'd use the previous model as the yardstick to dare us on what to do on the next model." Therefore, after the 1938 model year, Gregorie had to differentiate a Standard from a Deluxe only through different front-end treatments. The body shells could not be changed.

In this first year of distinct Standard and Deluxe models, Gregorie gained differentiation between the two by utilizing the sedan bodies from the 1937 models, designing two new front-end treatments—one for the Standard and one for the Deluxe—and designing completely new bodies for the Deluxe sedans, coupes, and open cars. By mixing and matching front-end treatments with different body designs, Gregorie provided cars with completely different appearances between the two models in all body types. The Standard model was available in the Tudor sedan, the Fordor sedan, and the coupe. The Standard sedans utilized the "hungry-looking" 1937 bodies, while the Standard coupe utilized the body design of the new coupe. The Deluxe models—the Tudor sedan, Fordor sedan, coupe, convertible, convertible sedan, and phaeton—all had new bodies and the new Deluxe front end treatment. (Figures 14-3 through 14-5)

Edsel's scheme was a mid-class car, similar to the General Motors Oldsmobile. For that, Edsel would have to wait until the 1939 model year.

To minimize tooling costs after the 1938 redesign, "Edsel seemed less receptive to an abrupt change," says Gregorie. "He developed a tendency to carry forward the basic effect of the previous

Figure 14-3. This is a three-quarter front view of the full-size clay model of the 1938 Ford sedan. The 1938 Ford was the first Ford designed by Gregorie in his new capacity as chief designer. (Photo from the Collections of Henry Ford Museum & Greenfield Village)

Figure 14-4. The pointy front end shown in this front view of the full-size clay model of the 1938 Ford sedan was Edsel Ford's preference. This effect would continue in Ford products for the next decade. (Photo from the Collections of Henry Ford Museum & Greenfield Village)

Figure 14-5. In this rear view of the full-size clay model of the 1938 Ford sedan, the crack in clay was caused either by inadequate support of the rear section of the model or because the model was moved. Thin bumpers, as shown on this model, were Edsel Ford's preference. (Photo from the Collections of Henry Ford Museum & Greenfield Village)

For the new Standard front end, Gregorie replaced the 1937 Zephyr-type grille design with a crisp, horizontal design that had grille bars running almost the full length of the hood. (Figures 14-6 and 14-7) "That spruced it up a bit," Gregorie says. "In other words, we tried to achieve a little more length-in-the-hood effect—to get away from that stubby effect of the previous model. The horizontal lines gave the front end a little more reach—gave it a more directional effect."

Gregorie designed the new Deluxe grille in a horizontal theme, similarly to the new Standard grille; however, the horizontal bars were laid out in a semicircular pattern and extended only a quarter of the way back along the hood. "The Deluxe grille had a considerable sweep to it," says Gregorie, "but by adding the 'soft nose' in this design, we lost a little hood length. Each grille is pleasing in its own way. It just depended on what your appetite was." (Figures 14-8 and 14-9) That is exactly what Edsel was seeking.

Edsel also wanted "a little more roundness" added to the new body design, so Gregorie eliminated the "bony" lines of the 1937 design and put some "meat" into the new Deluxe bodies by adding a longer crown to the front fenders, larger radius to the roof panel, and beautiful, sweeping lines to the rear end. Similarities existed between the carried-over 1937 bodies and the new 1938 bodies—the front fenders had a similar shape and were deep drawn, the rear fenders had a teardrop shape, and the windshield was of a two-piece, V-shape design. However, from a styling perspective, the new coupe, sedan, and open car bodies

Figure 14-6. This photograph shows the full-size clay model of the 1938 Ford Standard front end. Gregorie kept the deep-drawn front fenders of the 1937 model but emphasized the horizontal speed lines of the grille by extending them all the way back to the cowl. "That spruced it up a bit," explains Gregorie. "In other words, we tried to achieve a little more length-in-the-hood effect." (Photo from the Collections of Henry Ford Museum & Greenfield Village)

that Gregorie had designed were more graceful, fuller, and ten inches longer than the Briggs-designed 1937 bodies, although they were on the same wheelbase.

The 1937 Fordor and the 1938 Fordor "both have the same midship section," explains Gregorie, "yet the roof line is entirely different on the '38. It's domed and a little more rounded. (Figure 14-10) We also tried to get a little more flow to the rear end. Mr. Edsel

Figure 14-7. Shown here is another front-end treatment for the 1938 Ford Standard. (Photo from the Collections of Henry Ford Museum & Greenfield Village)

A similar distinction was achieved between the 1937 coupe and the 1938 coupe. (Figure 14-11) With the 1938 coupe, unlike the stubby 1937, says Gregorie, "we added a lot of graceful curves in the roof line and in the window shapes, and a nice sweep on the rear end." The two coupes have similar roof lines; that is, the door drip molding drops down at a sharp angle at the back of the roof and stops short of the belt line. This gives a very abrupt transition from the roof to the trunk lid. Nevertheless, the added rear-end length of Gregorie's 1938 coupe provided a streamlined, well-proportioned profile.

In Gregorie's opinion, "the '38 model was the transition from an old car to a new car. On the 1936 Ford, for example, the formation of the front end, the grille, and the sheet metal is more symbolic of an older car. On these cars, the headlamps were separate from the bodies, and the rear ends dropped off at a sharp angle. With the '38, you get an entirely new impression of the car. It has the headlamps built into the sheet metal, a more flowing roof line, and a different type of grille. Basically, it's the same car as the '36. It has the same wheelbase, and it works out the same from the wheel centers to the front of the grille. But those new front ends in the '38s gave a little more reach effect—a feeling of motion—leading to further changes in the next few years."

Ford liked what he called a 'fastback.' (I don't know how fast it had to go to look like that, but at least it gave that impression.) You'd be surprised how we were forced by constraint to keep the headroom in the back to thirty-eight inches from the seat—and we had to get those 'booshell baskets' in the trunk—and that controlled the roof line to a great extent. That's another reason we made it a fastback. If we made the roof line too flat, we would've had to move the rear seat forward, and that would have cheated the legroom."

Gregorie continued his plan to redesign the formerly stubby Ford into a better-proportioned, more flowing automobile in the 1939 models. For example, in the 1938 coupe, the deck lid line falls inside the rear fender contour, making the rear end appear short. (Figure 14-11) In contrast, in the 1939 coupe, Gregorie

Figure 14-8. Here is a production version of Gregorie's 1938 Ford Deluxe front end, as exemplified on this handsome convertible. (Photo from the Collections of Henry Ford Museum & Greenfield Village)

Figure 14-9. This was one of a selection of grilles that Gregorie designed for the 1938 Ford. In all such proposals, Gregorie was trying to achieve the effect of a longer hood. (Photo from the Collections of Henry Ford Museum & Greenfield Village)

Figure 14-10. Compare the profile of the production 1937 Ford sedan (top), designed by Briggs, to the profile of the production 1938 Ford sedan (bottom), designed by Gregorie. "The '37 was kind of a bad situation," explains Gregorie. "There was terrific pressure from the manufacturing people to take weight out of the car, and the only thing that the Briggs people could come up with was to make the body shorter and stubbier, thereby using less sheet metal. In the end, I think they were able to take about one hundred and fifty pounds out of the '37. It was a stubby little car—it didn't smack of luxury, that's for sure. With our '38, we were getting a more modern profile. It had more shapely fenders, and it was a fastback. The trouble with both of these cars is that the rear seat passenger is right over the rear axle, the chassis is extremely high off the ground, and the floor is high off the ground. And in order to get a thirty-eight-inch headroom, it gave you a high profile roof line. Both of these cars show the effect of these handicaps." (Photos from the Collections of Henry Ford Museum & Greenfield Village)

Figure 14-11. Compare the beautiful profile of Gregorie's 1938 coupe design (top) with Briggs's stark moribund 1937 coupe design (bottom). Gregorie's design is sweeping and better proportioned—a major consideration in Gregorie's mind—than the truncated approach followed by Briggs. The 1938 Ford had the same wheelbase as the 1937 Ford, but Gregorie was able to give the 1938 Fords a more substantial, streamlined appearance by adding a longer hood and adding more length to the body in the rear. (Photos from the Collections of Henry Ford Museum & Greenfield Village)

put a reverse curve in the roof's drip molding and extended it to intersect with the deck lid opening. (Figure 14-1) The deck lid then trails out to where it blends with the rear fender profile. "You've got to have the headroom, and it blends nicely to the deck lid," explains Gregorie. "The rear end was pleasing. It had a nice tail—quite a sweep. And the rear fenders had quite a graceful, long sweep to them as well." These changes made the back end of the coupe appear more flowing and longer, although the 1938 coupe and 1939 coupe are identical in overall length.

Another change that enhanced the design of the 1939 models over the 1938 models was the location of the headlights. "The '38 headlamp was placed in the valley between the fender and the hood side panel, an effect carried over from the '37," explains the designer. "On the '39, we repositioned the headlamp to the center of the crown of the fender, making a flush profile."

For all their enhanced beauty, the 1938 Ford line continued to suffer from the high frame, foreshortened front end, and "that hell of a draw on that front fender stamping." Gregorie could not do anything about the length of the front end, but he could redesign the front fenders to eliminate the deep stampings and fulfill Edsel's promise to Sorensen that the fenders would be easier to manufacture. As with so many of Gregorie's designs, what began as a manufacturing requirement ended as a styling sensation.

Relying on his mechanical aptitude, Gregorie quickly realized that the easiest way to eliminate the deep-drawn front fenders would be to reduce the distance from the outside edge of the fender to the inside edge. When he reduced the width of the front fenders in this manner, the front portion of the hood became wider and the grille opening became shorter and wider. After making several dozen sketches, Gregorie developed some attractive designs that would satisfy Sorensen's demands.

When Edsel launched the Standard and the Deluxe models, "there was an effort to keep the hood stamping the same and the grille opening the same," explains Gregorie, "and in some cases the actual grille openings were interchangeable—the grille insert that went into each car fitted the same basic stamping. We were able to get a different front-end design by simply changing the grille. Some of these dressmakers and some of these suit designers, they do the same thing: change the design enough to give it a new effect." To illustrate this process, the 1937 grille was changed into the grille for the 1938 Standard, and a new grille was designed for the 1938 Deluxe. In 1938, the Standard and Deluxe front ends were changed again, making what appeared to be new models for 1939. Finally, the 1939 Deluxe became the 1940 Standard, and an entirely new grille was developed for the 1940 Deluxe.

The front ends of all these Fords had pointy grilles and pointy hoods because "this type of front end appealed to Edsel," explains Gregorie. "Edsel liked pointy shapes." This design trend started with the 1936 Zephyr and was carried over to the 1937 Ford, both of which had an inverted boat prow effect. On the 1939 and 1940 models, Gregorie made the grille shorter, wider, and less sharp, but the streamlined, "cut-through-water" effect remained apparent. (Figures 14-12 through 14-25)

The softening of these front ends was a result of Edsel's desire to maximize the interchangeability of the various front-end parts among models. Gregorie notes that "the grille opening in regard to the fender shape between the '40 Standard and the '40 Deluxe is the same. The hood stampings, however, are different. On the Deluxe, the line stamped out along the belt line and the applied chrome—the belt molding strip—you get the impression of a longer hood, even though the hood is about the same length. The eye catches the belt molding immediately. It projects the hood and gives it a lead effect. On the Standard front end, it's sort of a blah—it all blends into the front end. On

Figure 14-12. This is an early rendition of the 1939 Ford Deluxe grille, built on 1938 Ford sheet metal. The ferociousness of this design was toned down in the production design, although the profile from the hood line downward to the bottom of the grille was similar. (Photo from the Collections of Henry Ford Museum & Greenfield Village)

Figure 14-13. These are early front-end ideas for the 1939 Ford. Placing one type of grille design on one side and another design on the other enabled Gregorie to develop various ideas with minimal work. "I don't think much of the double grille," explains Gregorie. "And the other one looks kind of heavy and coarse. It would have been good for a pickup." (Photo from the Collections of Henry Ford Museum & Greenfield Village)

Figure 14-14. In this early version of the 1939 Ford Deluxe grille, the "eyebrow" over the top of the grille is an interesting feature but never made it to production. The thin cross section of the hood later was extended downward to the crown of the fender. (Photo from the Collections of Henry Ford Museum & Greenfield Village)

Figure 14-15. The front end of the 1940 Ford Deluxe begins to take shape in the design department. In this design, Gregorie extended the hood as far past the front axle as practical to achieve a "more length-in-the-hood effect." (Photo from the Collections of Henry Ford Museum & Greenfield Village)

Figure 14-16. In this refined version of the 1940 Ford Deluxe front end, the center piece of the grille and its two outer wings are close to what went into production, but the hood opening remains under consideration. (Photo from the Collections of Henry Ford Museum & Greenfield Village)

Figure 14-17. On this pre-production version of the 1940 Ford Deluxe, the belt molding extends all the way to the nose of the hood. On the production model, the belt molding stopped a few inches from the nose. (Photo from the Collections of Henry Ford Museum & Greenfield Village)

Figure 14-18. This is a full-size clay model of the 1940 Lincoln-Zephyr three-window coupe. (Photo from the Collections of Henry Ford Museum & Greenfield Village)

Figure 14-19. This is a full-size clay model of the 1940 Lincoln-Zephyr Club coupe. (Photo from the Collections of Henry Ford Museum & Greenfield Village)

Figure 14-20. This photograph shows a full-size clay model that can be called the "missing link" between Gregorie's late-1939 Ford and Mercury designs and his early-1940 designs. It incorporates both Edsel's penchant for sharp, pointy shapes and Gregorie's preference for "full, round shapes." The prowl-like front end is similar to what eventually went into production as the 1939 Mercury, whereas the rear end resembles that of the 1941 Ford business coupe. "This is a proposal for a '39 Mercury business coupe," explains Gregorie. "The rear fenders look like those used on the new [1939] Mercury, but we never had a Mercury coupe like this for '39. There's more of a hump in the rear of this design than there was on the production Mercury. It looks a little too long in the rear for the length of the front; a little out of balance. There were all kinds of 'mud pies' we were making during this period—this is just one of them." (Photo from the Collections of Henry Ford Museum & Greenfield Village)

Figure 14-21. This photograph shows a full-size clay model of the 1938 cab-over-engine truck. "Not much styling went into commercial vehicles," says Gregorie. "They were controlled by a basic set of dimensions such as body width and wheel clearances." (Photo from the Collections of Henry Ford Museum & Greenfield Village)

Figure 14-22. Here is a full-size clay model of a proposed 1938 Ford delivery van. (Photo from the Collections of Henry Ford Museum & Greenfield Village)

Figure 14-23. This photograph shows a full-size clay design buck of a 1938 Ford pickup front end. (Photo from the Collections of Henry Ford Museum & Greenfield Village)

Figure 14-24. This is a clay model of the 1939 9N Ford tractor. "This tractor could make a story in and of itself," explains Gregorie. "The problem was, was it going to be called the Ferguson tractor or the Ford tractor with the Ferguson System? The debate kept going back and forth between Henry Ford and Harry Ferguson. To hedge our bet, we had two sets of name plates made for the clay model. When the old man [Henry] came up to see it, why, we put the Ford name plate on it, and when Ferguson came in, we put the Ferguson name on it. I had one of the boys ready with both name plates, and he'd come running out with the correct name plate when we saw one of them coming down the hall. It was a real circus. It ended up being called the Ford Tractor with the Ferguson System. That's the kind of stuff we used to go through." Nevertheless, Gregorie's Art Deco tractor design continued in production for thirteen years. In the "9N" designator, the "9" stood for the year of production (1939) and the "N" was the Ford designator for tractor. (Photo from the Collections of Henry Ford Museum & Greenfield Village)

Figure 14-25. This bus was designed in 1938 by Gregorie and his crew but was never meant for wide-scale production. It was designed to transport dealers and other visitors from the Ford Rotunda visitors center to the Rouge plant, administration building, and other Ford plants in the Dearborn area. (Photo from the Collections of Henry Ford Museum & Greenfield Village)

the Deluxe, it's a little more sharp. You get the feeling of motion in this front end, whereas the Standard looks more stationary. With the Deluxe grille, it's a little more fussy looking. It shows up better, reads better. It gives the impression that you get a little more chrome for your money. 'It gives better feeling for the eye,' as Dick Beneicke used to say."

Which front end is better?

"I'd say it's a tossup," exclaims Gregorie. "We had to have a difference between the two cars, and that's what we had in mind. The '40 Deluxe grille has a shoulder treatment on the side that blends back into the side panel, and that was enough to change the identity. The '39 Deluxe grille had a little bit of the flavor of the previous grille, and we went along with that theme for another year. We were controlled on the width and depth of the grille by the confluence of the catwalk line, which came down and formed the bottom of the hood, and the edges of the fenders. There were about sixty-nine different grille shapes that we could have used to fit that space, which amounted to just simply changing the texture of the grille, or peaking it up, or something of that sort. But we submitted to Edsel four or five different grille designs that fit the opening, and he made his decision from them. We knew the certain shapes Edsel favored, and we knew these types of grilles would be okay with him. They were harmless looking. They never created any discussion."

That was fine with Gregorie. He had neither the time nor the manpower to constantly develop front-end treatments. This was a business, after all, and, as Gregorie says, "the production people were at our heels to give them something to work on. But Edsel was very liberal. As long as it looked reasonable, as long as it looked nicely balanced and filled the bill—why, he'd okay it. He'd ask me if I liked it, and I'd say I liked it fine. He'd say okay, and we'd throw the design over the wall to engineering so that they could begin tooling."

The 1938, 1939, and 1940 Ford products were the last years in which there was a definite appearance of separate, distinct fenders. After that, the bodies grew wider, enveloping the fenders into the body side panels. "There's more definition or separation between the fender and the body on these cars," explains Gregorie. "This is apparent in the front fenders, and it is obvious in the rear fenders, especially when viewing the car from the rear. From this perspective, you can see how the belt line flows down into the valley between the body and the fender. There's a pronounced offset at the belt line—the fenders are completely separate." (Figure 14-26)

Regardless of their beauty, the 1938, 1939, and 1940 Fords did not turn out as well as they could have if Henry Ford had given Gregorie a new, up-to-date chassis and had changed his demands for having to carry milk cans or "booshell baskets" in the trunk of the vehicles. Henry always insisted that there must be enough head room for a man and his top hat.

"We had more headroom in our cars than the other manufacturers," explains Tucker Madawick, "because the old man [Henry] insisted that we sell these cars to farmers who went to the county fair wearing bib overalls—and, of course, they always wore their hats in the car. Harley Earl, with the full support of the top man at General Motors, Alfred P. Sloan, never had this problem. Earl said, 'Let 'em take their hats off!' The difference between Ford and General Motors was unbelievable."

Gregorie agrees. "We always wanted to get the profile of the car lower," he laments, "but the chassis height controlled that, and we didn't have much say in that. It would have run afoul with Sheldrick and the others. I could see Sheldrick now. Whenever I even proposed a chassis change, his face would turn red—he'd almost have apoplexy! The stuff I had to put up with. Oh, boy!"

Figure 14-26. This rear view of a 1940 Ford sedan clearly shows the demarcation between body and fenders. This distinction became less apparent in the 1941 models and was completely eliminated in the 1949 models. (Photo from the Collections of Henry Ford Museum & Greenfield Village)

The chassis height affected more than the road clearance of these cars—it affected each one's entire shape. "The deep body side panels and high belt line on these cars made the windows look too shallow, especially on the sedans," explains Gregorie. "This effect occurred because the seats sat on top of the frame rails, and it was quite a distance from the top of the seat to the running board. Originally, that sill line at the bottom of the doors used to be up high, like on those old Model A's. (Figure 14-27) The seats were high on those cars, too, but the splash aprons cut up the bodies so they didn't look so deep. Then suddenly, in these later cars, the door panel dropped right down to the running board, so the body appears extremely deep. (Figure 14-28) In order to get a lower overall look, we had to make the windows shallow, which in turn made the roofs thick. If you made the glass area larger, you'd have thrown the car all out of proportion. All this was caused by the height of the chassis. I wish I could have lowered that platform three or four inches. That would have allowed us to shorten up the side panels and open up the window openings. That would have made all the difference in the world."

Another drawback of these cars was the stubbiness of the front ends. "They were pretty heavy, husky front ends for the wheelbase," explains Gregorie.

Figure 14-27. In this 1929 Model A Fordor sedan, the splash apron, located between the door sill and running board, divided the body horizontally and gave the body a thinner effect. When the splash aprons were removed in later Ford designs, the door sill dropped down to the running boards, giving the bodies very deep side panels. (Photo from the Collections of Henry Ford Museum & Greenfield Village)

Figure 14-28. This side view of a 1939 Ford Deluxe Fordor sedan clearly illustrates the deep bodies, narrow windows, and thick roofs of the Fords of the 1930s and 1940s. This combination of factors was caused primarily by the height of the chassis off the ground. The floor in these Fords was almost two feet off the ground! (Photo from the Collections of Henry Ford Museum & Greenfield Village)

"These front ends remind me of somebody whose collar is too tight—you can see he's choked! It looks as though you parked a little carelessly, hit the curb, and shoved the front axle back about ten inches. That's exactly the type of effect you get. Looking at these cars, you almost want to take the front axle, jack it up, and pull it ahead. It looks like it needs several inches more wheelbase, leaving the sheet metal alone. These cars really needed it. They call out for it. But there was nothing I could do."

What Gregorie *could* do was minimize the effect. He accomplished that by increasing the hood length forward of the front axle center line on the 1939 models and again on the 1940 models. "To get more length in the hood itself," says Gregorie, "it was straightened up and drawn out just about to the "nth" degree to emphasize its length. But there was a limitation as to how much overhang past the front axle we could handle. What we would like to have done was to have left the hood alone and pull the front axle forward and add a little more of the fender behind the front wheel. That would have given these cars better proportions."

One trick that Gregorie could have used to give the front fenders a feeling of longer length was to extend the rear part of the front fenders into the front door area, a design cue implemented by Harley Earl on some General Motors cars in 1941. "I considered that," Gregorie says, "but Edsel wouldn't go for it. He wanted the front door clean."

Nevertheless, the 1938, 1939, and 1940 Fords are among the most handsome designs from the Ford Motor Company, reflecting the creative synergy between Gregorie and Edsel. The cars sold well when they were introduced, making the dealers happy. Billy Hughson, the first Ford dealer, applauded the new designs at a dealers' meeting. "I have watched Edsel Ford grow from a boy to a real man and executive," Hughson acknowledged. "I can see in him the same high ideals of Mr. Henry Ford, only more streamlined. As you know, we have him [Edsel] to thank for the beautiful models that we dealers have the privilege of now selling."

Hughson also should have thanked E.T. Gregorie.

CHAPTER FIFTEEN

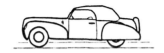

THE 1939 MERCURY
"Just a Stylized Ford"

Now that Edsel had the Standard and Deluxe Fords, the Lincoln-Zephyr, and the Lincoln to compete against some of the models offered by General Motors, he wanted a car that would enable a Ford owner to move up into the next price field, in the same way that a Chevrolet owner could move up to a Pontiac or Oldsmobile. In early 1937, Edsel began talking with John Davis and E.T. Gregorie about adding a new car to the Ford lineup—a car somewhere between the Ford Deluxe and the Lincoln-Zephyr.

Again, Edsel was putting the needs of the company before his personal dilemma. He knew even before talking with Gregorie and Davis that he would have to confront his father about the new project, and he was not looking forward to it. Nevertheless, he pressed onward.

To determine what size and price range the new car should be, Gregorie drew some charts that compared prices and weights of Ford products with the competition. The charts, Gregorie says, "plainly showed the hole in the structure that had to be filled.

There was about a $500 spread—which was a lot in those days—between the Ford and the Zephyr, and there was nothing in between."

Although hastily compiled, Gregorie's data reinforced what the three men had already concluded: that the new car should be slightly larger than the Ford and smaller than the Zephyr. Gregorie did not care about the size of the new car, but he certainly did not miss the opportunity to ask for a longer wheelbase. Edsel and Davis concurred, and the three men agreed on a wheelbase of 118 inches.

Finally, thought Gregorie, he would be able to work his magic on a suitable chassis, and he began sketching design ideas in earnest. Gregorie and Edsel never discussed the appearance of the new car because that was, after all, Gregorie's responsibility. From their preliminary discussions, Gregorie assumed that Edsel wanted the new car to appear unique, but Edsel vetoed every new idea from Gregorie. That was unusual because, until now, Edsel had been receptive to most of Gregorie's ideas. However, on the designs for this car, Edsel began steering Gregorie into designing the new car with styling cues taken from the latest Ford model, and this change in direction was baffling to Gregorie. He knew Edsel wanted the new car to be larger, and for that reason Edsel had approved the added length for the wheelbase, which Gregorie had wanted since the introduction of the 1935 models. From Gregorie's perspective, "Edsel wasn't trying to step it up far enough from the basic Ford. He didn't want a new design. He wanted a Ford-looking design!"

On the other hand, Gregorie believed the company needed a design that was entirely different from either the Ford or the Zephyr, and he continued to present those types of designs to Edsel. However, each time he did, Edsel became more irritated, and this conflict began to strain their relationship. "That was the thing that Edsel and I had a little funny feeling about," Gregorie tactfully explains.

Gregorie was bewildered. Why was Edsel coaxing him into designing a Ford-like car when the whole idea was to introduce an entirely new car? The answer finally came to light.

As Gregorie had surmised, Edsel wanted an entirely new car and thought that Gregorie's new designs were more than acceptable. The reason Edsel started insisting that the new design have Ford cues was because that is what Henry Ford wanted. As usual, Henry made Edsel fight dearly for the new mid-size entry. "The old man couldn't see why Edsel had to have two different cars with just a few inches difference in length and a different name," explains Gregorie. "Edsel was trying to sell the old man on this car as a complete break away from the Ford, but the old man was only going to accept it if it did smack of Ford. He [Henry] wanted to be able to point to the Ford insignia on the glass and on the battery box, and all that kind of stuff." Of course, Henry also wanted the new car to have his beloved chassis beneath it.

These requisites were all diversions. "Henry Ford was pretty cagey," says Gregorie. Henry simply did not want Edsel to add this car to the Ford product line any more than he wanted to add any of Edsel's continental cars, and he merely made these demands in hopes of killing the project. Henry knew that Edsel was hoping for a car with an entirely new appearance, and therefore he used that against Edsel. Henry felt that if he insisted on the car having a Ford identity rather than its own unique identity, Edsel would drop the project. However, Edsel called his father's bluff. Instead of abandoning the project, Edsel subtly began steering Gregorie into designing a Ford-like car.

Although Edsel never explicitly explained this dilemma to Gregorie, Gregorie noticed Edsel's subtleties while talking about the new car and finally put two and two together. "It was that kind of feeling in all my conversations with Edsel. I could see that he was up against a problem, and I had to go along with it. To have a complete breakaway would have been more than we could have asked for, and I was in no position to poke a hole in the idea and start a whole new complication."

With Gregorie's desire to develop an entirely new look, he had to be careful about how he developed the new design, although he now knew what direction to take. "We just weren't in a position to show Edsel anything that looked un-Ford-like," he says. "There was always that pressure, and I had to be awfully careful not to offend Edsel and his case and wants in dealing with the old man. I had to talk out of three sides of my mouth. It's unbelievable what I went through. I had to keep Edsel happy, and he had to keep his old man happy. The old man was smart enough to realize that I was trying to influence Edsel in a new direction, and I could see that Edsel was caught in the middle. He would come to me and in essence say, 'Dad can't see it this way, and I'm having a hell of a time satisfying him.' He couldn't put it in as many words, but he did express this fact to me very clearly. I was beat going home some nights, after spending hours with Edsel trying to work this thing out."

The design Gregorie eventually developed "was enough of a change so that it didn't look *exactly* like a Ford," Gregorie says. "Of course, I would have liked to have done something to give the new model an entirely different identity. But because of the pressures that were exerted in the development of the car, Edsel felt safer going along with his father, and I had to go along with it."

The appearance of the new 1939 Mercury followed the general theme of the 1939 Ford, except the body shape was a little wider and a little fatter. "In other words," describes Gregorie, "it was just a stylized Ford. We added a little more bulk to the car, increased the hood length, and altered the rear end a little, and it showed up on the road as a little more important-looking." The rear fenders on the Mercury had a little less wheel opening than the Ford. "That was one of the features of the car," says Gregorie. "The sheet metal actually came down over the tire, making it look a little thick and heavy. We were trying to get more of a closed-wheel-cover effect, sort of like a fender skirt."

Gregorie also gave the Mercury a more slanted windshield, and in the interior, he incorporated his latest design idea—the two-spoke steering wheel. Until this time, Ford had used what was commonly termed "banjo" steering wheels, which had three sets of steel spokes connecting the center hub to the outer rim at the twelve, four, and eight o'clock positions. Gregorie called them bicycle spokes. "I thought, one day, to make a steering wheel that was simple in appearance—make it so that the driver could rest his hands on the spokes of the wheel, and still be able to look straight through the steering wheel to the instrument panel." So Gregorie went to Jim Lynch, the shop foreman, and showed him a sketch of what he wanted. "Turn up a damn ball about the size of a large grapefruit with the guts of a hub in it, and let's put a couple of tapered spokes in it," he told him. After Lynch had completed it, Gregorie installed the new steering wheel on a new 1938 Ford phaeton that he planned to take on vacation to Maryland. On the way, "I stopped at a couple of dealer friends of mine to get their opinions," Gregorie says. "I'd frequently stop and talk with dealers and get their opinions on this and that. 'Oh, boy,' they said, 'that's nice and smooth, and you can see through it!' Well, anyway, Edsel bought the idea, and we came out with that two-spoke steering wheel shortly thereafter."

Keeping the Mercury Ford-like was only one of two requirements imposed on Gregorie. The other was ease of manufacture. The chrome window frames in the doors—now thought of as a styling feature of the Mercury—were done primarily to eliminate the need for a full-size door stamping. "We cut the doors off at the belt line," explains Gregorie, "and used a chrome channel frame to run the glass up into. That gave the sedan coupe a kind of a semi-convertible effect, a hardtop look. But the reason it was done was to make the doors cheaper to produce. With this design, we didn't have to have a full-size door stamping for the coupe; we just used the lower stamping, which, by the way, was also used on the convertible."

If you compare the 1939 Ford coupe to the new 1939 Mercury coupe, subtle differences become apparent. "The Ford coupe had a reverse curve in the rear end," Gregorie explains. "It was a continuous sweep from the rear seat head room on down to where it connected with the deck lid. On the Mercury, the roof line broke off sharply, giving the appearance of a little more deck lid and a shortened-up roof. You can call the Ford a 'slowback' and the Mercury a fastback."

The most significant feature of the new Mercury, from Gregorie's point of view, was its longer front end. "You see, on the Mercury, you don't quite get that stubby appearance as on the Ford," he proudly explains. "The Mercury was basically the same design as the Ford, the same fender design, and so on, but the front end didn't look stubby as it did on the Ford. There was more fender between the front door and the front tire, and that makes a great improvement in the overall proportions of the car. When you put the Ford coupe (Chapter 14, Figure 14-1) and the Mercury coupe (Figure 15-1) face to face, you could see the ill effect of the stubby front end on the Ford."

Figure 15-1. The avant-garde 1939 Mercury coupe. Its ancestry is evident when compared to its cousin, the 1939 Ford (Chapter 14, Figure 14-1), but the Mercury had a four-inch longer wheelbase, the added length added forward of the cowl. This added length enabled Gregorie to give this car better proportions. The stainless steel around the door and quarter windows gives the car a hardtop effect, which was not intentional—the design allowed the coupe and convertible to use the same door stamping. (Photo from the Collections of Henry Ford Museum & Greenfield Village)

As far as the front end was concerned, "we didn't try to make it too different," says Gregorie. "We made several versions of it, but I could feel Edsel waltzing back to that typical Ford-type front end. He was partial to having a Ford identity, and he carried that as a club in dealing with the old man. He wanted enough Ford identity so he could justify this damn car along with the Ford. I'd like to have seen more variation there, but Edsel felt it safe to stick to that. Henry wanted as much Ford appearance as possible, which was not the right way to go about it, in my mind, but nevertheless that's the order we got." (Figure 15-2)

Bud Adams, working under the direction of Dick Beneicke, modeled the Mercury grille. Instead of copying the vertical grille bars of the Ford, Adams used a horizontal grille design with finely spaced bars placed at a sharp angle. This gave the Mercury a wider front-end appearance. "Dick Beneicke helped show me how to proportionally space the bars as they went from the bottom and swept up past the bumper," explained Adams. "The bars would not only be spaced wider as they went up, but the bars themselves were wider. So that was the tricky thing to make right, but when it was there, it was a real improvement, and it carried the flow lines of what you might call the 'catwalk,' or the area between the fender and the hood, right on forward and down behind the bumper." (Figures 15-3 through 15-6)

When the design of the new car had been decided, it was time to develop a name for the car. Again, Gregorie ran into trouble. Edsel had chosen the name Mercury from a list of more than one hundred other names, which included everything from Cyclops

Figure 15-2. If you compare this Mercury sedan with the Ford sedan (Chapter 14, Figure 14-28), the similarities and the differences between the two vehicles become apparent. It is evident that both cars come from the same "family," but Gregorie incorporated more rounded, bulbous forms in the new Mercury, while the Ford captures Edsel's desire for separate, pointy shapes. (Photo from the Collections of Henry Ford Museum & Greenfield Village)

Figure 15-3. In this early full-size clay rendition of the 1939 Mercury, notice that the hood stamping is very shallow, similar to the 1937 Ford or 1936 Zephyr. Also, bulges in the fenders give some added dimension to the headlights, and the grille has a horizontal appearance. (Photo from the Collections of Henry Ford Museum & Greenfield Village)

The 1939 Mercury

Figure 15-4. This photograph shows a three-quarter right-side view of an early clay model of the 1939 Mercury. The front end was completely changed on the production version, but the longer front fenders, more slanted windshield, more rounded body, and more enclosed wheel opening on rear fenders were all design cues incorporated into the final design. (Photo from the Collections of Henry Ford Museum & Greenfield Village)

Figure 15-5. A few months later, the Mercury front end has been changed. The hood has been thickened, bulges in the fenders have been eliminated, and the grille has been given a little more of a "V" shape. (Photo from the Collections of Henry Ford Museum & Greenfield Village)

Figure 15-6. This photograph shows a front view of a clay model of the 1939 Mercury nearing its final form. "I put those bulges along the sides of the hood to give a point of interest," says Gregorie. (Photo from the Collections of Henry Ford Museum & Greenfield Village)

to Winged Victory. The name itself did not cause the commotion; rather, it was Edsel's insistence in combining the name Mercury with the name Ford. Reminiscent of what he had done with the Lincoln-Zephyr, Edsel wanted to call the new car the Ford-Mercury. Gregorie was aghast. The design that was intended to be an entirely new car was already diluted to resemble a Ford, and by calling it Ford-Mercury, almost no chance existed to separate its identity from that of the lower-cost Ford.

Years earlier, Gregorie noted a similar contradiction with the Lincoln. He mentioned to Edsel that it was incompatible to sell and service Lincolns at Ford dealerships and suggested that the two lines be separated. "Can you imagine a man paying $6,000 for a custom Lincoln," he told Edsel, "and then having to take his luxurious motorcar into a Ford garage to get it serviced alongside an old dump truck or a tractor!" Edsel got a big kick out of that. Joe Galamb and the other chassis people would never think of that, because they were not sales oriented. However, Gregorie was, and Edsel appreciated Gregorie's sales perspective. Nonetheless, when Gregorie mentioned to Edsel that the Mercury name should be separated from the Ford name, Edsel responded heatedly, "What's wrong with the Ford name? We've spent forty years building up the Ford name, and you're telling me it's not good enough to put on this automobile!"

"Boy, that was like a hot poker on Mr. Ford," recalls Gregorie. "He kind of had me on the hot seat, squeezing it for a minute!" Nevertheless, Gregorie defended his argument. "There's nothing wrong with the Ford name, Mr. Ford," Gregorie quickly replied, "but we need to get people to believe it's not a Ford so that they will move up to it, just like General Motors gets people to move up from a Chevrolet to a Pontiac, and from a Pontiac to an Oldsmobile." Edsel left without saying anything further.

Gregorie now thought his days were numbered, but he pursued his instincts and had some of the sheet-metal "boys" hammer out some hubcaps with only the Mercury name on them. Meanwhile, the Ford-Mercury was shown at the New York Automobile Show, and many of the Ford district sales managers and dealers "climbed all over Edsel" when they saw the Ford name connected with the new Mercury. They explained to him the same thing that Gregorie had tried to explain: it would be a grave mistake to associate the Mercury name with the Ford name.

During the return train trip to Detroit, Edsel thought about what the dealers had told him, and by the time he reached Detroit, he changed his stance on naming the new car. It went into production as the Mercury (although the Ford logo appeared on the windows, the battery, and other parts), and Gregorie kept his job.

In the end, the Mercury "came out well enough," explains Gregorie. "In fact, we were tickled to death to get an okay on a car that was built as well as this one was." Tucker Madawick, one of Gregorie's apprentice designers at the time, agrees. "The Mercury was a real advanced car against the basic Ford," Madawick says. "The car was avant-garde. And to get something like that into the Ford line was a real accomplishment. Remember, the old man still had a say in everything that Edsel did, and that was a big problem. I hate to imagine what Edsel went through with old Henry Ford in getting it approved." (Figure 15-7)

Figure 15-7. This was the lineup of the Ford Motor Company for 1939 (left to right): Lincoln, Lincoln-Zephyr, Mercury, Ford Deluxe, and Ford Standard. When Gregorie became design chief, the company was offering only the Lincoln and Ford models. With his hard work, Edsel's support, and the acquiescence of Henry Ford, Gregorie managed to implement three additional model lines in only four years, fulfilling Edsel's dream of offering Ford models against most models that General Motors offered. (Photo from the Collections of Henry Ford Museum & Greenfield Village)

CHAPTER SIXTEEN

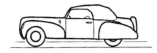

THE 1941-1948 FORDS AND MERCURYS
"That Hangdog Look"

For the 1941 model year, Gregorie and his design team completely redesigned the Ford and Mercury lines. It was the second time they had done so in three years. (Figure 16-1) Gone were the streamlined rear ends and the sharp, pointy front ends on the cars. "We had so much of that pointy effect in the late 1930s," explains Gregorie, "that by the time we went into another series of cars, which started in '41 and '42, we decided to pursue a fresh-looking effect—a fuller, rounded, more bloated and horizontal effect. I sold Mr. Ford on the horizontal effect." (Figures 16-2 and 16-3)

In particular, the 1941 Ford was the transition car. Its front end (Figure 16-4) combined portions of the pointy grille of the 1940 Deluxe Ford (Chapter 14, Figure 14-17) with portions of the soon-to-be completely horizontal grille design of the 1942 Ford (Figure 16-19). As with its 1937 predecessor, the 1941 Ford was a unique, one-year design, and customers either loved it or hated it.

The wheelbases of the Ford and the Mercury were increased two inches each in 1941, with the Mercury maintaining its four-inch lead over the Ford. The 114-inch wheelbase of the Ford and the 118-inch wheelbase of the Mercury subsequently remained unchanged throughout Gregorie's tenure.

Starting with the 1941 model and carrying through to the 1948 model, Gregorie says, "All the body surfaces were flatter and blended in one contour." That is, he softened the lines of the 1941 models, which began assuming a more modern appearance. As Gregorie explains, "We started to thin the roof down, we added more glass area, and we added a little more curvature, a little more bulge, to the side panels—they weren't as flat as in the previous models. We wanted to reduce the bony look of the earlier models, so we gave them the feeling of a more bulbous effect." These cues are most apparent on the sedans. "At the trunk edge opening, the body is fatter. The body looks to be pulled in and more tumblehome to the roof, and the body bulges at the belt line. At the fender valley, in

Figure 16-1. E.T. Gregorie poses outside his office in 1940 in front of his latest creation, a 1941 Ford. Gregorie was at the zenith of his design career at this time, and during the next six years, he would revamp the entire Ford, Mercury, and Lincoln lines twice! (Photo from the Collections of Henry Ford Museum & Greenfield Village)

between the body side panel and the fender, there's no definition of a valley at all. It's all one surface. The fender blends into the deck lid. It's more rounded and bloated." (Chapter 14, Figure 14-26; Figure 16-5).

More definition or separation existed between the fender and the body on the 1940 model than on the 1941 model. "The surfaces come together on the '41," Gregorie continues. "The '41 was a much smoother design. The fender blended right into the sheet metal of the deck lid, and the quarter panel also blended in. In other words, the belt line ends at the deck lid opening, whereas on the '40, the belt line flowed down into the valley between the body side panel and the fender. On the '41, the rear fender was a much easier stamping to make; the depth was very shallow. The body contours just flow right into the deck lid, and the deck lid forms the back end. There was a pronounced offset at the belt-line level on the '40; the belt line just faded out at the drip molding on the '41—it all just faded into the back panel. The lines were more horizontal on the '41, which gave it a faster appearance. The belt line, instead of drooping down with the deck lid, trailed out with the forward motion of the car, and the fender contours and surfaces tended to wrap around the back, whereas with the '40, the fenders were completely separate."

Unlike the 1938, 1939, and 1940 sedans, which had graceful fastbacks, the 1941 models had a much rounder appearance. Gregorie whimsically describes the 1941 sedan rear end as a *monkey rump*. "See, it looks like the back of a monkey as it's hunched over, perched in a tree!" he exclaims. (Figure 16-6) "It was smart looking. You couldn't go wrong on that kind of roof shape. Edsel would go for that every time. You didn't have much leeway to work in. There wasn't much you could do with it. You're given an overall length, and so much

The 1941–1948 Fords and Mercurys

Figure 16-2. The lower, wider, and fuller effect of the 1941 Ford design (top) is evident when compared to the design of the previous model (bottom). (Photos from the Collections of Henry Ford Museum & Greenfield Village)

Figure 16-3. These are three-quarter rear views of the 1940 Ford sedan (top) and the 1941 Ford sedan (bottom). Both are fastback designs, but the belt line on the 1940 model flowed down into the valley between the body side panel and the fender, whereas on the 1941 model, the belt line trailed outward with the forward motion of the car. (Photos from the Collections of Henry Ford Museum & Greenfield Village)

Figure 16-4. This photograph shows the front end of the production 1941 Ford. (Photo from the Collections of Henry Ford Museum & Greenfield Village)

Figure 16-5. Compare the rear view of the 1941 Ford sedan shown here with the rear view of the 1940 Ford sedan (Chapter 14, Figure 14-26). Gregorie made significant changes between the 1940 and 1941 model years. He made the 1941 models wider by bringing the body side panels outward over the fenders, thus eliminating the effect of a separate fender on the 1940 models. The distance between the outside of the fenders was the same on the 1940 and 1941 models; therefore, the interior width of the 1941 was wider than that of the 1940. Gregorie also eliminated the split rear window used in Ford products since 1937 and designed a simple oval-shaped, one-piece rear glass as exemplified in this rear view of the 1941 Ford sedan. (Photo from the Collections of Henry Ford Museum & Greenfield Village)

Figure 16-6. In this profile of the 1941 Ford sedan, Gregorie refers to the back-end profile of this design as a "monkey rump." The 1941 models were the first to receive completely new body designs since the 1938 models. "We were trying to get a fuller effect, add more roundness to the body surfaces," explains Gregorie. "The bottom of the doors roll out over the running boards—we did that on the Zephyr's [doors] as well." The rear end, a fastback similar to the previous sedans, was more rounded, and the profile line appeared to tuck under at the bottom. "We were required to have that terrific amount of head room right over the rear axle, and there wasn't much we could do about it. It wasn't until engineering moved the engine forward and the rear passenger containment forward that we were able to lower the profile." (Photo from the Collections of Henry Ford Museum & Greenfield Village)

overhang from the rear wheels, from the bottom of the trunk lid out to the bumper, and you had head room limitations. And then you had to get those 'booshell baskets' in the trunk. Well, you were pretty well hemmed in. It was about as pleasing a roof line as you could evolve. It served its purpose in terms of appearance."

The 1941 coupe, especially the business coupe (Figure 16-2, top), was a striking departure from the 1940 design (Figure 16-2, bottom). With the 1941 model, Gregorie extended the front-end overhang as much as possible to balance the front end with the rear end. In retrospect, Gregorie now prefers the design of the 1941 model because it gives the impression of speed. "That's a pretty nice-looking car," Gregorie says. "There's a little more depth to the deck lid, not as streamlined as in the '40. It's humped up a little more, providing more room for the 'booshell baskets.' That was an important issue! We were supposed to take that seriously!"

Gregorie cannot say as much for the front end of the 1941 model. "The '41 was not a very credible design," he says apologetically. "It looks a little droopy. It has that hangdog look. It's a down-in-the-mouth-looking grille, kind of a chinless effect." That was the result of Charley Sorensen altering the design at the last minute to make the front fenders easier to manufacture.

Bud Adams is credited with designing the front end of the 1941 Ford. As Adams explained it, the full-size clay models for the Ford and Mercury were almost completed except for the front ends, and he was assigned to develop a front end for the Ford model. Usually, apprentices such as Adams were given a sketch or a drawing from either Gregorie or one of the design managers—Kruke, Wagner, or Walter—from which to work. In this case, Adams was given no direction and was left to his own devices. "I knew the legal heights for the headlights," Adams explained, "and I knew where the bumper should be, and what acceptable overhang on the front would be, and I knew where to find the corners of the radiator. So I went to work, like a crazy beaver, and modeled this damn front-end design directly in the full-size clay model. A few mornings later, our beady-eyed, mustachioed Bob Gregorie came by, did a double-take, returned to the clay model, and with a smile said, 'Hey, that's not bad!' " (Figures 16-7 through 16-10)

With that encouragement, Adams put the final touches on his clay model. As he was adding aluminum foil for the plated parts of the grille, he detected a movement of people behind him. "I turned around, and here's the whole top echelon of the company—Charley Sorensen, Pete Martin, John Crawford, Larry Sheldrick, and Al Wibel—marching in to take a look at this 1941 Ford front end. (Figure 16-11) They wanted to see what they were going to have to do to build this thing."

Figure 16-7. In this early design proposal for the 1941 Ford, the vertical grille piece is similar to what eventually went into production, but the two mesh-covered openings on either side of the vertical piece were replaced with two oval-shaped grille pieces. Note the odd-shaped joint line between the inner and outer portions of the front fenders. This was only one idea that Gregorie and Bill Pioch, head of production tooling at Ford, developed to comply with Charley Sorensen's directive of making the front fenders easier to manufacture. (Photo from the Collections of Henry Ford Museum & Greenfield Village)

Figure 16-8. Except for the front-end treatment, this design of what eventually became the 1941 Ford business coupe nears completion. Note the similarity between this body design and that of Gregorie's Mercury design shown in Figure 14-20 of Chapter 14, which was done more than two years earlier. (Photo from the Collections of Henry Ford Museum & Greenfield Village)

Figure 16-9. This photograph shows the full-size clay model of the final design of the 1941 Ford. Eugene "Bud" Adams is credited for the design; Charley Sorensen is credited for ruining it. (Photo from the Collections of Henry Ford Museum & Greenfield Village)

Figure 16-10. This photograph shows the full-size clay model of the 1941 Ford sedan coupe. Gregorie extended the roof line on this coupe vis-à-vis the business coupe and abruptly terminated the roof drip molding at the belt line, similarly to what he had done on the 1939 Mercury. He also added the horizontal crease in the rear fender to match the fender separation line in the front fender created by Charley Sorensen. (Photo from the Collections of Henry Ford Museum & Greenfield Village)

Figure 16-11. These were the engineering and production executives of the Ford Motor Company in 1941. Standing (left to right): Dale Roeder, truck engineering; Larry Sheldrick, Ford engineering; William Pioch, tooling; Jack Wharam, Lincoln engineering; E.T. Gregorie, design chief; unknown; Ed Scott, body engineering; and Gene Farkus, design engineer–engines. Seated (left to right): Frank Johnson, Lincoln engineering; Pete Martin, vice president–production; Charles Sorensen, vice-president–manufacturing; and Joe Galamb, body engineering. This is the last known photograph of the old guard—the group of men who had started early with Ford Motor Company and were instrumental in its phenomenal success. Within three years after the photograph was taken, Martin was dead, Sheldrick and Sorensen had been fired, and Johnson had retired. Likewise, Gregorie was fired in 1944, but he was recruited back by Henry Ford II a year or so later. One of the conditions on which Gregorie insisted before returning was that Joe Galamb be kept at bay. Henry Ford II readily complied by picking up the phone and having Galamb fired. (Photo courtesy of E.T. Gregorie)

Although the appearance of the front end was pleasing, the rookie Adams made the same fatal mistake that had been made in the 1937 and 1938 Ford front ends. The front fenders on his design had too much of a draw from the outside edge to the inside edge, and that immediately caught Sorensen's attention. He would not allow Adams' design to go into production the way it was. "No way!" Sorensen said, as he walked around the front of the clay model. "No way!"

As Adams rose from his stool and backed out of the way, Sorensen picked up the designer's clay modeling knife and started drawing parting lines in the fender. First, he carved a horizontal line that ran from the back of the fender up to the center of the headlight, and then he drew a vertical line directly above the center of the wheel opening. Within a few minutes, Sorensen had transformed the one-piece, deep-drawn fender into a shallow-drawn, three-piece fender, ruining the design in the process.

The front fenders did not go into production exactly as Sorensen had cut them, but they were a close facsimile. Bill Pioch, one of Sorensen's men and the chief tool designer at Ford, discussed ideas with Gregorie about how best to make the deep-drawn fenders easier to manufacture, without necessarily following Sorensen's suggestions to the letter. "We changed it around two or three different ways," explains Gregorie, "and finally settled on splitting the fenders right in the middle, just under the headlamps. It was split all the way around, and we added a chrome molding [on the Super Deluxe models] to cover up the connection."

Those front fender separation lines "got to be a headache," says Gregorie. "There was no way to appropriately connect the profile separation line with the front end. There was a hump in the profile, which the parting line followed, but then it dropped off right under the headlamp, and you picked up a new styling theme by the time you got around to the grille. It was never very well connected. Actually, I added the rear fender crease because of the separation line that developed in the front fender. I picked up the sweep of the belt line, and it made quite an effective treatment."

Because the Mercury was on a longer wheelbase, it had a longer hood and did not have the same fenders as the Ford; however, the fender stampings were split the same way as those on the Ford. Instead of having a three-piece grille as did the Ford, the Mercury utilized a horizontal-shaped two-piece grille, which "looks a little neater than the Ford," says Gregorie. (Figures 16-12 and 16-13)

Figure 16-12. In this front-end design proposal for the 1941 Mercury, the top part of the grille would have an insert, and the lower portion would simply be stamped into the sheet metal. In the production design, the grille piece was carried downward to the bottom portion of the sheet metal stamping. (Photo from the Collections of Henry Ford Museum & Greenfield Village)

Figure 16-13. This full-sized clay model of the 1941 Mercury nears its final form. The louvers in the sheet metal below the grille were not part of the production design. (Photo from the Collections of Henry Ford Museum & Greenfield Village)

During the 1941 model year, Edsel and Gregorie introduced a third model in the Ford line, the Super Deluxe. The Special (formerly called the Standard), Deluxe, and Super Deluxe all had the same basic sheet metal, "but we got a three-way identity with added trim," explains Gregorie. Added stainless steel trim, different paint schemes, and different interior trim and fabrics differentiated one level from the other. All these features were added, says Gregorie, "just for a showroom effect."

Although the 1941 line began as attractive and new, Sorensen's manufacturing concerns turned the new line into something less desirable. As Gregorie summarizes it, "We were glad when that year was over with, so we could move on to better things."

On the 1942 models, Gregorie made the full transition from a vertical grille to a horizontal grille on both the Mercury and the Fords. A hint of a vertical piece in the center of the Fords remained, but it was merely a continuation of the vertical bars laid out in a horizontal theme. (Figures 16-14 through 16-19) On the other hand, all the grille bars on the Mercury were laid out horizontally. (Figures 16-20 and 16-21) The Fords had a full frame around the grille, whereas the grille on the Mercury has no frame at all. "The Mercury grille was kind of a stiff, square-looking

Figure 16-14. This was a unique front-end design proposal for the 1942 Ford. "If we had gone along further, we might have done something with this!" exclaims Gregorie. (Photo from the Collections of Henry Ford Museum & Greenfield Village)

shape," describes Gregorie. "It had a double horizontal molding in the fenders. We did that on the postwar Mercury as well. When you compare the '42 to the '41, the '42 looks much better, crisper. That '42 is quite effective, more modern-looking. It looked like we were heading off in the right direction. I hate to think of the pains we went through in order to achieve it!"

Figure 16-15. Here is another design proposal for the front end of the 1942 Ford. This design, and the design shown in Figure 16-16, "were companion efforts," explains Gregorie. (Photo from the Collections of Henry Ford Museum & Greenfield Village)

Figure 16-16. In this design proposal, the 1942 Ford was created on a 1941 sedan body. "That's a coarse-looking brute," explains Gregorie. "We were hitting bottom in this design. But gradually we'd whittle it down and refine it and develop something that was acceptable for production." (Photo from the Collections of Henry Ford Museum & Greenfield Village)

Figure 16-17. This photograph depicts the early throws of the 1942 Ford front end. "This grille kind of looks like the later Mercury," explains Gregorie. "The grille disappears on the top and the bottom. It's just a cylindrical roll with an opening on top and an opening on the bottom." (Photo from the Collections of Henry Ford Museum & Greenfield Village)

Figure 16-18. The bold, horizontal grille for the 1942 Ford begins to take shape on a 1941 Ford body. "We were still looking for something to put in that grille opening," explains Gregorie, but this is close to what eventually went into production. Note the trim ideas for the right headlight, park lights, and hood. The convex moldings around the nose of the hood did not go into production. (Photo from the Collections of Henry Ford Museum & Greenfield Village)

Figure 16-19. Here is the final clay mock-up of the 1942 Ford front-end design. By this year, Gregorie had made the full transition from vertical front grille treatments to horizontal grille treatments. (Photo from the Collections of Henry Ford Museum & Greenfield Village)

The 1941–1948 Fords and Mercurys

Figure 16-20. The profile of this 1942 Ford sedan coupe is similar to that of the 1941 Ford sedan coupe shown in Figure 16-10; however, on the 1942 model, Gregorie added an ogee transition (a reverse curve) from the roof panel to the trunk lid, which gave the rear end a more flowing effect. This body style was continued through 1948. (Photo from the Collections of Henry Ford Museum & Greenfield Village)

Figure 16-21. This is the final version of the 1942 Mercury. (Photo from the Collections of Henry Ford Museum & Greenfield Village)

Figure 16-22. The lines were simple but effective for the front-end treatment of the 1946 Ford line. (Photo from the Collections of Henry Ford Museum & Greenfield Village)

"The '41 body carried right on to '48," Gregorie says. "What you had was a slot to put a grille in, and that controlled the whole thing. The cars had the same hood and the same fenders as the prewar models. I remember I sketched up that grille for the '46 model one morning about ten o'clock. The production men were after us to get going with something—and I wanted to get home to lunch—so I grabbed a piece of paper, made a rough sketch, and said to my men, 'Make this thing fit the damn opening, and let's go.'" He retained the upper part of the full frame of the 1942 model and filled the remainder with three thick horizontal bars trimmed with stainless steel. (Figure 16-22) The 1946 models had a red stripe painted in a recess in the stainless strips, whereas the 1947 and 1948 models had

When the United States entered World War II, all production of civilian vehicles ceased. The next new Ford models were the 1946 ones, which Ford started producing in the late summer of 1945. To begin production as soon as possible, the 1946 models were nothing more than facelifted 1942 models. Unknown to Gregorie at the time, the 1946, 1947, and 1948 Fords were to be the last ones to carry his imprint.

a smooth finish. "It was all right," says Gregorie about his grille design. "It was acceptable. It had a very nice balance. The grille had a little more rigidity to it. We just had to get something fresh. The grille effect was very strong and purposeful-looking."

For the postwar Mercurys, Gregorie designed a bold die-cast grille with a heavy appearance. It was similar to the grille of the 1942

Ford; however, the grille bars on the Mercury were spaced closer together, making the grille appear almost solid. (Figures 16-23 and 16-24) "It was enough of a difference to change the identity," explains Gregorie, "and that's basically what we were after. The hoods on these cars were straightened out and pushed forward just about to the limit to balance the rest of the car. This was particularly evident in the business coupe. The hood was stretched out enough to counterbalance that overhang in the rear." (Figure 16-25)

Gregorie adopted the horizontal theme on the 1942 Lincoln line also. As on the facelifted Fords and Mercurys, all Lincoln bodies for 1942 remained basically the same as they had been in previous years, but Gregorie replaced their famous "butterfly" grille with a horizontal grille that had a more substantial appearance. (Figures 16-26 and 16-27) This new front-end treatment was incorporated in all 1942 Lincolns, but Gregorie illustrates its development by describing the Continental.

Figure 16-23. This photograph depicts a front-end proposal for the 1946 Mercury. (Photo from the Collections of Henry Ford Museum & Greenfield Village)

Figure 16-24. Here is the final version of the 1946 Mercury. (Photo from the Collections of Henry Ford Museum & Greenfield Village)

"Just before the war," Gregorie recalls, "we designed the heavier-looking front end for the 1942 Continental. All the General Motors cars had been beefed up by this time—especially the Cadillac with the new 60-Special—so they looked heavier and a little more important on the road. Compared to these cars, the Continental—with its razorback hood, those skinny fenders, and little, skinny bumpers—looked like a hungry horse. Edsel, of course, liked those features, but I finally convinced him that a car in its price range has to look a little more important on the road. So that's when we developed the new hood, new front fenders, and horizontal grille. Once we did that, though, the body began to look a little skinny. We never changed the body. The doors, the windshield, and the floor pan were all the same. With its big, husky-looking front end, it looked a little out of proportion in places, like so many of those facelift deals. But it was still a right decent-looking car. The '42 front end was a nice-looking front

Figure 16-25. This is the profile of the production 1946 Ford business coupe. (Photo from the Collections of Henry Ford Museum & Greenfield Village)

Figure 16-26. The horizontal grille treatment on this clay mock-up is close to what finally went into production on all Lincoln, Lincoln-Zephyr, and Lincoln Continental cars in 1942. Gregorie likes this version of the Continental better than his first. (Photo from the Collections of Henry Ford Museum & Greenfield Village)

Figure 16-27. This photograph shows the full-size clay model under development for the mid-1940s Lincoln, circa 1942. (Photo from the Collections of Henry Ford Museum & Greenfield Village)

end. The horizontal bars were very nice. It looked important, anyway. Unfortunately, there were no new mechanical features connected with the updated version. If we could have introduced a proper engine at that time—a good straight-eight engine, for example—it would have been perfect. The '42 Continental, by the way, was the last decision Edsel made on the Lincoln."

After the war, Gregorie made the last changes in his most famous design—the Lincoln Continental—but the result was not as successful as he would have hoped. "The Lincoln people wanted something flashy, something that would compete against the massive front end of the Cadillac," he says. "So we came up with this massive-looking grille for the '46 Continental, which I didn't care much for. It got away from us. It was a rush-rush job, so we just pushed it through. It looked like the grating for a furnace! (Figure 16-28) The '42 had a better-looking grille. The new grille was really dragging that body down. My goodness, that body was developed in 1936, and here we're up to '48! Wow!

Figure 16-28. E.T. Gregorie poses next to his personal 1947 Lincoln Continental in the fall of 1946. Gregorie went to this massive front-end treatment on the Continental from the gracefulness of the earlier model "to make it look more important on the road," he says. Gregorie always liked to dress up his personal cars to give them a unique appearance. Although this car appears to be stock, it had a special gray paint job, red leather interior, dual exhaust, and a special spring-spoked steering wheel. On the left side can be seen specially turned large chrome-plated wheel covers; on the right side, the large wheel covers were painted the body color. (Photo courtesy of E.T. Gregorie)

Yet, oddly enough, the '46 to '48 Lincolns were very popular with their massive grilles."

One of Gregorie's more extraordinary Ford designs was a postwar model that was developed, again, from unrelated events. Since the Model T days, Ford produced wooden-bodied vehicles. The Model T versions were called depot hacks; however, in 1928, with the coming of the Model A's, they were called station wagons. Starting with the 1938 models, Gregorie and his men did all the design work on these wooden-bodied station wagons. "We made full-size drawings of these bodies on a blackboard, and the body draft was made in Dearborn," explains Gregorie. "The bodies, however, were built at Ford's Iron Mountain plant in Michigan's upper peninsula, and shipped down in freight cars and mounted in Dearborn."

Although the bodies were made completely of wood, "they held up remarkably well," says Gregorie. "I always owned one and got a new one every couple of years. As long as you kept plenty of varnish on them, they were fine. A curious thing, though—the roof was covered with a type of oil cloth over wooden slats, and if you didn't keep the wood well varnished and kept the weather out from under the oil cloth, why, dampness would get in the roof cage, and then you'd get a crop of mushrooms around the roof rail on the inside of the car!"

Until 1946, no manufacturer had made a wooden-bodied convertible. Dubbed the Sportsman, Gregorie's wooden-bodied convertible was both unique and stunning. The whole idea started accidentally when Henry Ford II (president of the company in 1946 because his father, Edsel, had died in 1943) noticed a complete Model A chassis in the corner of Gregorie's design studio. It had been there since the early 1930s so that Henry Ford could reminisce about the good old days.

Reminded of the times Edsel and Gregorie worked together building special cars, Henry II asked Gregorie to design a body for the chassis. "I'd like to have a little car to use down at the beach at Southampton," Ford told Gregorie. "Can you do me up some kind of a little beach wagon, something to use down there, take the kids to the beach?" So Gregorie sketched a body that was part station wagon and part convertible, and had it made from wood at the old aircraft terminal, the same place he had made Edsel's special sports cars a decade earlier. (Figure 16-29)

During the war, the Ford aircraft plant made gliders from beautiful mahogany plywood, and Gregorie used this expensive material to build the little car. An interesting feature of the car (harking back to the Model Y) was its one-piece hood that came all the way back to the windshield. "It was all hand built," says Gregorie. "It also had a special floor pan and a folding khaki top. But the Model A frame was so light (it was like a couple of bed rails) that when the tailgate was dropped down and somebody sat on it, the doors would pop open! Fortunately, we found that out before we turned it over to Number Two Ford [that is how Gregorie sometimes referred to Henry Ford II], and we reinforced the frame. It rode rougher than hell."

As Henry II was picking up the little car, Gregorie, as bold as ever, said, "Mr. Ford, when you get through with this little thing, I'd like to acquire it." In the same way as his father, young Henry II was more than happy to oblige. That fall, after the kids had used the car all summer, Henry II sent the car to the design department. "It so happened," recalls Gregorie, "that the very day that he sent his driver out there with it to my office, there was a group standing around—John Bugas [the company's manager of industrial relations], Ernie Breech [the company's new executive vice president], and two or three others—and we all got to talking about this little car. Breech said, 'I've got to have this, I've got to have this. Oh boy, I've got to have this.' I said, 'I'm sorry, Mr. Breech, it's already mine. I spoke with Mr. Ford about obtaining it several months ago. His driver is just sending it out for me so I can take it home.' Breech never forgave me for that."

Figure 16-29. This little Model A beach wagon was designed by Gregorie for Henry Ford II and was made from a Model A chassis that had been sitting in the corner of the design department since 1931. It precipitated the production of the Sportsman wooden-bodied convertible. (Photo from the Collections of Henry Ford Museum & Greenfield Village)

Unlike so many of the other special cars that Gregorie obtained from the company, he owned this little makeshift Model A for quite a while. "I had it for six or seven years," he says. "I had that car at Grosse Ile for a while, and then I brought it down to Florida when I retired. When my wife, Evie, and I started living on our boat, it finally became a problem having to rent a garage to keep it in. (You couldn't leave it outside. You had to keep it varnished.) I tried trading it in for a new Ford station wagon, explaining to the dealer that it used to be Mr. Henry Ford's car, and that it was the only one ever built like this—it probably cost $50,000 to build. You could say that it was the last Model A the company ever built. Finally, I sold it to an automobile collector for fifteen hundred dollars."

"That little Model A wagon was the inspiration to wood-panel a convertible," continues Gregorie. (Figure 16-30) "We made a handmade one in the winter of '45 or the first of '46. We

Figure 16-30. In Ross Cousins' renderings of wooden-bodied convertibles, the top vehicle shown here is one of the initial ideas for a production model, and the bottom vehicle exemplifies the production Sportsman. (Top photo from the Collections of Henry Ford Museum & Greenfield Village; bottom photo courtesy of Ross Cousins)

sent it down to Florida so that Number Two Ford could use it. Instead of sending it back to Dearborn after Henry's vacation, I had it taken up to the Jacksonville branch office, where I picked it up in April 1946. Evie and I drove it out to Fort Lauderdale and then up to Dearborn."

Gregorie is proud of his postwar models, although they were only facelifted designs. "That front end was pretty snappy," he says. "The overall appearance, it looked ready to go!" When he compares the 1940 coupe to the 1946 version, Gregorie prefers the later model. "The '46 coupe design was more rakish-looking—looks anxious to go—as compared to the '40 coupe. Those speed lines—the horizontal line treatment on the fenders—gave the '46 coupe a feeling of motion."

By 1948, the body styles developed by Gregorie almost a decade earlier were beginning to appear somewhat tired and outdated. "That was an awful stretch!" Gregorie says. "But it's surprising how well that old model looked in '48, the last of this run of cars. Actually, those carry-over cars started after the war were pretty decent-looking, considering what we had to work with."

CHAPTER SEVENTEEN

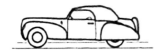

POSTWAR DEVELOPMENT
"A Little Ford and a Big Ford"

When America entered World War II in December 1941, many of Gregorie's design staff left the department. Some were drafted, others joined the service on their own, and others were reassigned to the Ford Willow Run bomber plant near Ypsilanti. Tucker Madawick recalls that "Bill Wagner chose to go into the Ford B-24 bomber program, working in the engineering department at Willow Run. I committed myself to B-24 inspection, hoping for pre-flight and flight inspection at the same facility." Gregorie and a few of his men received government deferments to work on bomber turret designs, camouflage, and military vehicles. However, during breaks in work on these projects, Gregorie and his small staff managed to develop completely new Ford, Mercury, and Lincoln models.

The body lines Gregorie developed for the 1941 Fords and Mercurys had more than run their course by 1948. "Those bodies," Gregorie says, "went back to the Dark Ages." Edsel also realized this, and early in 1941 "he sort of made up his mind that we needed pretty drastic changes." Again, Edsel met with John Davis and Gregorie, and he began talking to them about developing an entirely new line of automobiles. These were not to be facelifted models with the same old chassis and drive train with which Edsel and Gregorie had been strapped for the past decade; rather, they would be completely new cars with new bodies, chassis, and drive trains. During their discussions, Gregorie also suggested the idea of adding a small car to the new Ford lineup. Edsel was more than receptive to that idea because he too was thinking about a small car. Edsel had told Joe Galamb in passing only a short time earlier that "we will have to build a cheaper car someday."

Henry Ford was elderly by the beginning of World War II, and Edsel had seized more control of the production end of the business and was overseeing the construction of the Willow Run bomber plant as well. (According to government officials who oversaw bomber production, the men in control at Ford at this time were Edsel Ford and Charles Sorensen. These officials considered Henry Ford merely a figurehead of the company.) Now, with more power than he had ever held in the past, Edsel felt the time was ripe to pursue a new line of cars.

Over the years, the idea of producing a small-size Ford was discussed occasionally among the Ford executives, and the need for it by the early 1940s seemed greater than ever. (Figure 17-1) One hindrance that had prevented the company from introducing a smaller car was a suitable engine.

Since 1932 when old Henry began installing V-8 engines exclusively in all Ford products, the competition had proclaimed that the engine was an uneconomical power plant, a criticism the public took seriously during the Depression. To counteract this criticism, the company, at Edsel's insistence, developed a little 60-horsepower V-8 engine and made it an option in 1937. The intent was to maintain Henry's integrity by keeping the new engine a V-8 but to reduce its size, thus increasing its fuel economy. The little V-8 offered better gas mileage, but "it wasn't powerful enough to get out of its own way," says Gregorie, especially when the engine was installed in a full-size Ford. Edsel understood this dilemma and again convinced his father in 1939 of the need for a practical six-cylinder engine, something similar to the powerful and fuel-efficient six-cylinder engine from Chevrolet. Two years later, Ford made available an optional six-cylinder engine.

The new flat-head six was a well-made engine, "a good lugging engine," as Gregorie describes it, and it got good gas mileage. However, even this engine was too large for the size of car Edsel

Figure 17-1. Shown here is a proposal for a small Ford on a shorter wheelbase, created by Gregorie and his design team in 1939. This car was never produced, but design cues used in this design eventually were incorporated into full-size Fords in 1940 and 1941: the two-piece vertical grille similar to the grille used on the 1940 Ford, and the "monkey rump" rear end incorporated in the 1941 through 1948 Ford sedans. This car also was to have a six-cylinder engine, something Edsel had been wanting for years. (Photos from the Collections of Henry Ford Museum & Greenfield Village)

and the others were contemplating. For a short time, the small V-8 was considered for the small car, but that idea was soon discarded in favor of a new, smaller six-cylinder engine.

Gene Farkus, the Hungarian-educated engineer, was asked to design a small six-cylinder engine for the new small Ford that Gregorie would design. Farkus was the expert engine designer at Ford and was instrumental in developing the Ford tractor engine,

Postwar Development

Figure 17-1. (continued)

the Model A engine, and several experimental engines for Henry Ford. During the development of the Model A but prior to turning his energy toward the development of a V-8 engine, Henry hoped to outshine Chevrolet with an X-8 engine, which Farkus designed and built. When that engine proved to be unreliable, Henry remained with the dependable four-cylinder configuration for the Model A and then began working on a low-cost V-8.

The latest engine on which Henry Ford had Farkus work was a five-cylinder engine. Henry had several of these engines built and installed in production Fords for test purposes. One day, Gregorie and his jokester friend Walter Kruke obtained one of these experimental cars and took it to lunch. The engine ran well at high speeds, but it idled poorly. Seeing an opportunity to have some fun, Gregorie and Kruke pulled into a gas station and complained to the mechanic that the engine of their car was misfiring. The mechanic opened the hood, took one look at the engine, and walked to the side of the car. "Hey," the mechanic said, "you guys know that your car only has five cylinders?"

"Five cylinders!" exclaimed Gregorie and Kruke in unison. "We'd better take this back to the dealer and have him give us a new car!"

The five-cylinder engine never went further than these few experimental prototypes.

Meanwhile, Gregorie and Edsel decided to pursue their new line of cars, which included a small Ford and a big Ford, and Edsel instructed Farkus to keep working on a new six-cylinder engine for the small Ford. "Everybody felt pretty good about the idea of two sized Fords," Gregorie recalls. "At that time, cars were increasing in size, weight, and price—the trend was generally upward. In other words, we had pushed the basic Ford up to a higher level, and we wanted to fill in beneath it with a small Ford. That was the way the thinking went. We were certainly headed for a small car in some manner or form, and we felt we were on pretty safe ground by building a small car."

Thus, the decision was made. The new small Ford would be designed around a 98-inch wheelbase, while the larger Ford would be built on a 118-inch wheelbase. According to Edsel's directions, Gregorie was to make the small Ford and the big Ford appear "as similar as possible." (Figures 17-2 and 17-3)

With the decision of the small Ford completed, Edsel, Davis, and Gregorie finalized the other models in the lineup. In addition to the small Ford and the big Ford, there would be a new Mercury and two new Lincolns, a small one and a big one which, as with the Fords, would be similar in appearance. (Figure 17-4) All models would be available in a variety body styles: a coupe, a convertible, a Tudor sedan, and a Fordor sedan. The exception was the Lincolns, which would not be offered in a coupe. The big convertible Lincoln also would be offered as a Continental by adding the trademark trunk-mounted spare tire. (Figure 17-5) "Between the two of us," explains Gregorie, "Edsel and I decided on the proper shape of these new cars. We thought as one about what these cars should look like. In the meantime, engineering was supposed to be paralleling our work as far as the chassis, suspension, and various other mechanical features were concerned."

Figure 17-2. This is Gregorie's second attempt at developing a smaller Ford design after the small Ford project was resurrected by Edsel Ford, John Davis, and Gregorie. Proposed names for the little car were "Scout" and "Pacer." (Photos from the Collections of Henry Ford Museum & Greenfield Village)

Figure 17-3. These renderings by Ross Cousins show a small Ford based on Gregorie's clay model. At the top is a Tudor sedan; at the bottom is a station wagon. (Photo courtesy of Ross Cousins)

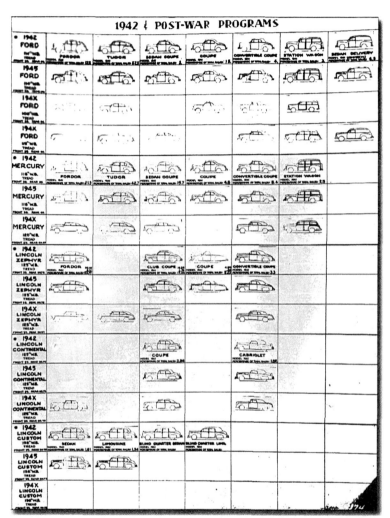

Figure 17-4. As design chief, Gregorie also assumed many responsibilities that product planners normally handle in the automotive industry today. Gregorie and Ed Martin compiled this chart to show Edsel the product plans for the company for 1942 and as soon as World War II ended. (Chart from the Collections of Henry Ford Museum & Greenfield Village)

Figure 17-5. This photograph shows the full-size clay model of the proposed postwar Lincoln Continental. "This is as close as we came to a Continental," says Gregorie. "You can see the spare tire mounted on the back end of it." Unfortunately, the Continental never went into production as part of Ford's postwar lineup. "Edsel had not approved this design," continues Gregorie, "and we didn't feel that it was distinctive enough to offer as a Continental." The Continental name plate would not return until 1956. (Photo from the Collections of Henry Ford Museum & Greenfield Village)

Al Esper, a mechanic and test driver, recalled a heated discussion among the executives in 1941 about changing the chassis on the new cars. "Sorensen, Martin, Mr. Edsel Ford, and Mr. Henry Ford were all in the engineering lab," Esper explained. "Quite a battle was going on as to the design of these cars and front spring suspension. Mr. Edsel Ford was fighting the battle for the coil spring suspension. Sorensen and Martin were for the coil springs as well, but they didn't have enough gumption to back up Edsel Ford. So, naturally, he lost. They didn't want to be crossed up by Mr. Henry Ford."

Nevertheless, Edsel wanted a new chassis, and he instructed Sheldrick to work surreptitiously on various types of independent front suspension systems for the new cars. Several designs were developed, and a few months later, Gregorie recalls driving one Ford prototype which had independent torsion bar suspension.

Designing a small Ford had brought Gregorie full circle. His first assignment, back in 1931, had been to design a body for a short-wheelbase chassis. Now, a decade later, he was to do that again. However, this time he was practically free of constraints—no chassis restrictions and no width restrictions. Gregorie now also had free reign over body design. For the first time in his career, Gregorie did not have to contend with some extraneous factors.

Gregorie began by developing design ideas for the little Ford/big Ford combination. He made fastback designs, trunk-back designs, and convertibles for both sizes. For the little Ford, he toyed with both rear-wheel-drive and front-wheel-drive designs. Although front-wheel drive did not take hold in the United States until the late 1970s, at this early date Gregorie already appreciated its space-saving capabilities for small cars, and he had several models made to test its viability for the new small Ford. "For the front-wheel-drive car," Gregorie explains, "one of the possibilities we considered was to place Farkus's new engine transversely into the chassis, with the transmission along side of it with stub axles coming out each side. It made a nice compact picture. We also explored the possibility of standing the engine upright to save space. You know, placing the engine with the flywheel-end down. It would have saved all kinds of space. We built a wooden model of it, just to have it in position, something to look at." Although a unique idea for an automobile, the arrangement was not new. According to Gregorie, General Motors had built a vertical-mounted diesel engine for submarine chasers during World War II, and that is what gave him the idea. "They stood about six feet high and developed about one thousand horsepower," Gregorie recalls. "They worked well, but it made a strange-looking engine room, I can tell you that!"

To keep Henry Ford from upsetting the whole project, "the front-wheel-drive concept was done mostly on the Q.T.," recalls Gregorie. They did not necessarily hide their work, but they did not place it in public display either. Ordinarily, Gene Farkus worked on engine development in the engineering lab, but Edsel instructed him to work on the small six-cylinder engine at the Rouge plant, away from the view of old Henry.

Larry Sheldrick, Gregorie's long-time nemesis, was fully aware of the front-wheel-drive scheme and reluctantly went along with it. However, Sheldrick was aggravated more by the fact that Gregorie was working on mechanical aspects of the new cars that historically were Sheldrick's responsibility. "I guess it would have aggravated us just as well if Sheldrick and his men started making body shapes," confesses Gregorie. "But I knew that Edsel was interested in mechanical innovations of various kinds, and we didn't attempt to go over anyone's head. We thought we might as well see what we could do. As a matter of fact, because of the war work the company was doing for the government, the chassis people weren't supposed to do development work anyway. But we could!" (Figure 17-6)

By November 1941, Gregorie and his design team had full-size clay models of the two proposed Ford sizes. (Figures 17-7 through 17-9) Both were fastback designs, and because Gregorie was not concerned about putting "booshell baskets" in the trunk, the roof lines were lower and the sloping rear ends were not as extreme as

Figure 17-6. This is a design buck of Gregorie's front-wheel-drive "little Ford." The gentleman in the back seat is Martin Regitko. (Photo from the Collections of Henry Ford Museum & Greenfield Village)

Figure 17-7. The design of this early full-size clay model of the "little Ford" sedan is much more refined than Gregorie's earlier idea, as shown in Figure 17-2. (Photo from the Collections of Henry Ford Museum & Greenfield Village)

Figure 17-8. This is an early full-size clay model of Gregorie's "big Ford" Tudor design, completed in February 1943. Note the similarity in the appearance of this car to the appearance of the "little Ford" shown in Figure 17-7. Edsel Ford wanted a similar appearance between the big Ford and the small Ford, and that is what Gregorie provided. Edsel had approved this fastback design months earlier, but he also gave Gregorie approval to develop a trunk-back design to complement it. This particular model called for a stamped metal grille with applied stainless steel strips; however, as with all designs in this early stage of the design process, this part was subject to change, and it usually did. (Photo from the Collections of Henry Ford Museum & Greenfield Village)

they had been on the 1938 through 1941 models. Edsel favored fastback designs, so that is what Gregorie developed. "Yeah, it was all right with me," Gregorie says. "Whatever pleased Edsel. As long as it looked reasonable, why, I didn't favor one or the other, fastback or notch-back. Both of them basically fulfilled what we had to work with."

For years, Edsel and Gregorie had wanted to lower, widen, and reduce the overall height of their cars. With their new design program, those were the attributes toward which they worked. (Figures 17-10 through 17-13) "We were looking for something that was lower and wider—within reason," says Gregorie. "In other words, get the body profile as low as possible, consistent

Figure 17-9. This photograph shows a full-size clay model of the "big Ford" Fordor sedan. (Photo from the Collections of Henry Ford Museum & Greenfield Village)

Figure 17-10. This rendering, done by one of Gregorie's designers in February 1941, demonstrates the progressive thinking at Ford, even at this early date. The window area, or greenhouse, was more expansive, and the body was longer and lower to the ground. (Photo from the Collections of Henry Ford Museum & Greenfield Village)

with adequate headroom." This led Gregorie to what he refers to as the slab-side look: side body panels brought outward past the plane of the wheels, fenders melding into the body, a lower roof height, and a more extensive "greenhouse" area. "By this time," continues Gregorie, "we were losing the benefit of fender identity, fender shapes, which are wonderful. They formed wonderful relief in those older cars. But when we got into this slab-sided thing, the body looked like a loaf of bread. You had to do things to relieve the surface, to give the car identity, and to relieve the ponderous bulk of it. Originally, all my new designs had the front fenders blending into the doors. I remember getting my clay knife (I used to carry one with me all the time) and putting a dip in the door belt line. (Figure 17-14) It was a pretty line. It tapered gracefully, and it dropped down—a kind of sparky little drop. I was trying to bring back the feeling of a fender and to give the cars some identity. If you stripped this off—as was done initially—there was nothing. It looked just like a balloon."

Overall, the little Ford/big Ford designs were lower to the ground—something for which Gregorie had been arguing almost since he had started working for the company. Also, the belt lines were lower, the roofs were thinner, and the glass area was more extensive. Following Edsel's directions to the letter, Gregorie made both cars appear almost identical, except for their size, which shows his genius. It is almost impossible to make a small car look as good as a big one, especially when the intent is to make the two sizes appear similar, but that is exactly what Gregorie achieved.

"I designed the little Ford and the big Ford with fastbacks because Edsel was partial to fastbacks," explains Gregorie. "But once he had approved these two designs, the intent was to develop a business coupe and a convertible for each Ford size, as well as a bustle-back or trunk-back design for both sedans. Ever since we came out with the Continental, with its trunk-back design, Edsel slowly took a liking to that concept in our other cars."

Figure 17-11. This photograph shows the quarter-size model of the proposed 1943 Ford. Early in 1941, both Edsel Ford and Gregorie agreed that a new line of Ford automobiles needed to be designed for introduction in 1943 or 1944. This is one design proposal that was considered. (Photo from the Collections of Henry Ford Museum & Greenfield Village)

Figure 17-12. Regarding this quarter-size clay model of a future Ford, Gregorie comments, "I don't like the looks of this one. It has a short, stubby hood, and great big, long tail; it looks overloaded." (Photo from the Collections of Henry Ford Museum & Greenfield Village)

Figure 17-13. This photograph shows a quarter-size clay model of a mid-1940s Ford proposal. "One of my design apprentices worked on this," says Gregorie. "It's a good thing he stopped when he did. I don't care for that big, thick molding running down the middle of the doors." (Photo from the Collections of Henry Ford Museum & Greenfield Village)

Figure 17-14. The design shown in this early full-size clay model of the "little Ford" coupe has design cues from both Gregorie's new concept of wider, lower cars and pre-war lines, as well as the little dip in the door belt line—a Gregorie trademark. The front end is based on Gregorie's new direction, whereas the rear end is reminiscent of the 1942 Ford coupe. "The deck lid mimics the roof panel," explains Gregorie. "There's enough offset in the deck lid stamping to give it character. There's no attempt to keep it flush with the fender—it has kind of a positive offset all the way around. That drip molding, as it comes down around the quarter glass and down to the trunk, that's about the only way it could be handled; otherwise, it would go off into limbo." (Photo from the Collections of Henry Ford Museum & Greenfield Village)

Figure 17-15. In this full-size clay model of the "little Ford" convertible, the center upright in the grille of the model is reminiscent of the 1941 Fords. "This car didn't get very far," says Gregorie. (Photo from the Collections of Henry Ford Museum & Greenfield Village)

Edsel, Davis, and Gregorie also settled on the design for the remainder of the new Ford line. Similar to the two Fords, Gregorie was instructed to develop Tudor and Fordor sedans, coupes, and convertibles for the Mercury and Lincoln lines. He and his men started all these as soon as Edsel approved the basic design of the two Ford sedans. With this lineup of vehicles, they felt they could successfully compete on any level with General Motors or Chrysler in the postwar market.

"The most important point of this entire project," emphasizes Gregorie, "is that Edsel Ford himself approved this entire lineup of new cars. It was a very bold move on his part, and he took it very seriously. It wasn't just a presentation, as such; he was about to approve an entirely new line of Ford cars that had absolutely no input from his father! He gave it a hell of a lot of thought. No one's ever given the man credit for making such a momentous decision."

At any other automobile company at this time, it was the president's prerogative to make such decisions. However, this was not true at Ford Motor Company. Until now, most of Edsel's decisions were either overruled or nullified by his father. This time, that would not happen, not on this new lineup. Edsel realized that his father was at a point where he could offer little resistance, and Edsel acted quickly, approving this handsome array of cars developed by Gregorie for the postwar market. Production tooling began almost immediately. (Figures 17-15 through 17-29)

Unfortunately, this new lineup of cars was the last to be anointed by Edsel Ford. Shortly after meeting with Davis and Gregorie, Edsel became gravely ill. He had complained about his stomach

Figure 17-16. In this full-size clay model of the "little Ford" coupe, circa June 1944, the ogee line in the side panel of the model is a precursor to that used on the postwar Mercury. (Photo from the Collections of Henry Ford Museum & Greenfield Village)

Figure 17-17. Here is a full-size wooden model of a proposed "little Ford" station wagon. (Photo from the Collections of Henry Ford Museum & Greenfield Village)

Figure 17-18. This photograph shows a full-size clay model of the "little Ford" coupe. (Photo from the Collections of Henry Ford Museum & Greenfield Village)

Figure 17-19. This is a close-up of Gregorie's "little Ford" sedan. "This is a stamped grille," explains Gregorie. "We got the body shape all finished, but I don't recall a selection for a grille. This was one possibility." Note the clay model of the "big Ford" coupe in the background. (Photo from the Collections of Henry Ford Museum & Greenfield Village)

Figure 17-20. After Edsel and Gregorie settled on a design direction for an entire new line of automobiles in 1941, Gregorie began some design proposals for consideration. Well along by April 1942, this is an early fastback/boattail design for the "big Ford." (Photo from the Collections of Henry Ford Museum & Greenfield Village)

Figure 17-21. In this full-size clay model of Gregorie's postwar "big Ford" sedan, the body shape has been finalized, but the grille remains subject to change. This grille was only one proposal. (Photo from the Collections of Henry Ford Museum & Greenfield Village)

Figure 17-22. Here is another version of the "big Ford" sedan. (Photo from the Collections of Henry Ford Museum & Greenfield Village)

Figure 17-23. This photograph shows a quarter-size clay model of the "big Ford" convertible. "This double-louvered grille didn't correlate very well with the shape of the sheet-metal work above it," explains Gregorie. "The rolled grille I designed for it later tied the sheet metal and grille better." (Photo from the Collections of Henry Ford Museum & Greenfield Village)

Figure 17-24. This is a quarter-size clay model of the "big Ford" station wagon. A design similar to this went into production in 1949 as the Mercury. In the production version, the stainless steel molding shown running along the length of the model ran only from the front of the car to the front door. (Photo from the Collections of Henry Ford Museum & Greenfield Village)

Figure 17-25. This photograph depicts a quarter-size clay model of the "big Ford" sedan delivery. A standard body type since the Model T days, the sedan delivery was dropped for the 1949 model year because of cost constraints. However, it was revived in 1953. (Photo from the Collections of Henry Ford Museum & Greenfield Village)

Figure 17-26. This is a full-size clay model of a postwar Lincoln, circa March 1943. (Photo from the Collections of Henry Ford Museum & Greenfield Village)

for years, but his doctors had misdiagnosed his illness simply as ulcers—a malady that beset many men at Ford. They even had a word for it: *Forditis*. Edsel ate ice to deaden the pain.

In January 1942, his condition took a turn for the worse. "I had known of Edsel's troubles for some time," Sorensen later explained, "and I was frightened to see the pain he was in at times. On one of our trips to Washington, D.C. [during the war], he was suddenly taken ill after a seafood dinner. I got him back to our hotel rooms and called a doctor. Edsel was in such agony that I sat up all night, afraid that he was dying." In February, Edsel entered the Henry Ford Hospital, where surgeons removed part of his stomach. After the operation, the doctors sent him to his Florida retreat in Hobe Sound to rest. They restricted Edsel to a special diet of low-fat foods, water, and crackers, and he seemed to recuperate fine. Robert Rankin, Clara Ford's chauffeur and Edsel's friend, saw Edsel in Florida and said that he "looked wonderful" and that he had never seen "a man with such a fine suntan." However, Rankin soon learned that Edsel was not doing as well on the inside.

Coincidentally, Rankin had had a similar operation a few months earlier, and Edsel asked him what he was eating. "Oh, anything!" exclaimed Rankin. "French-fried potatoes, steak, fish."

Figure 17-27. Although Gregorie is best known for his automobile designs, he also was responsible for the commercial line from Ford. Here is a quarter-size clay model of the 1948 Ford half-ton pickup. This design was the beginning of the long-running F-Series line of commercial vehicles from Ford. This handsome line was a vast improvement over the homely 1942 through 1947 pickups. (Photo from the Collections of Henry Ford Museum & Greenfield Village)

"That's the trouble," complained Edsel. "They don't give me anything to eat. They just feed me pills!"

"You can't go far on pills," said Rankin.

"No," chuckled Edsel, "I get hungry once in a while."

"Why don't you come down to Palm Beach and have dinner with us?" asked Rankin.

"I wish I could, but I can't," Edsel replied. "Doctor's orders."

When Edsel returned to Dearborn in the late spring, he took on his normal workload as if nothing were wrong. He went to his office in the morning, met with his father and other Ford officials for lunch, and visited Gregorie in the afternoon. Albert Lepine, Edsel's secretary for twenty-five years, did not notice anything different in his boss. "I had no idea, seeing him almost every day at the office, that he was seriously ill," Lepine recalled. However, Edsel was gravely ill, and in November, he suffered a relapse. This time, Dr. Roscoe Graham, who was described as "one of the most noted abdominal surgeons of the Western Hemisphere," came to Detroit from Toronto to operate on Edsel.

Figure 17-28. This commercial vehicle, a 1948 panel delivery, went into production as shown in this full-size clay model. (Photo from the Collections of Henry Ford Museum & Greenfield Village)

Figure 17-29. This photo was taken in June 1945 and shows Gregorie's complete Ford lineup for the postwar market. In the left row are the Mercury (front) and Lincoln (back). In the middle row is Gregorie's "little Ford" lineup: in front is the sedan, followed by a one-tenth model of a convertible, ending with a coupe. In the row to the right are the "big Fords": a fastback sedan in front, and a trunk-back sedan in the rear, each showing different grille designs. On the right side of the display floor are renderings of various Ford body styles and ideas for the commercial line. (Photo from the Collections of Henry Ford Museum & Greenfield Village)

Finding cancer, Dr. Graham told Edsel's wife, Eleanor, "I'm sorry, but I can do nothing to help Mr. Ford." He gave Edsel less than a year to live.

Edsel was bedridden in the hospital for approximately a month and then was taken to his Gaukler Pointe mansion, where he remained through January 1943. In February, Edsel was well enough to travel to his winter home in Florida, where he stayed until early March. (Figure 17-30)

Throughout this period, Gregorie had no contact with his patron but remained at work developing postwar designs for the Mercury and Lincoln. "I had quite a lot of stuff ready for Edsel to approve," he says, "and I went a couple of months trying to contact him." In the old days, Gregorie would mail design ideas to Edsel at his Florida retreat, and Edsel would return them with his comments, thereby keeping Gregorie on schedule. However, this time, Gregorie did not know where Edsel was. "I called Mr. Lepine several times a week, asking when Edsel would be coming in, but he was very evasive. Finally, he told me that Mr. Ford had gone down to Hobe Sound and that he couldn't be disturbed."

When Edsel returned, he unexpectedly walked into Gregorie's office through the back door. "I could see that he appeared very ill," recalls Gregorie. "He had a beautiful tan, but he had lost a lot of weight." They talked for a while, Edsel asking in a general way how things were going. They never addressed any specific design or project, and they didn't walk through the design department. Finally, in a moment of compassion, Gregorie offered some advice to Edsel. "Why don't you go over to the eastern shore of Maryland and buy a nice, big farm over there, and just go away," Gregorie told Edsel. "Spend all the damn time you want there. This thing will hold together."

"I wish I could," Edsel solemnly replied. "I wish I could."

"Edsel didn't see any of the additional models we had worked up for him," says Gregorie. "He just came in that one afternoon, sat down for about an hour, then left. I never saw him again."

Gregorie's patron, the man who had started him on his long trek in automobile design, died early in the morning on May 26, 1943. He was only forty-nine years old. Edsel's death surprised everyone. Many people knew Edsel was ill, but they did not know the seriousness of his condition. Ross Cousins had talked to Henry Ford the day before Edsel's death and asked about Edsel's health. "He [Henry] said he was on his way to visit him [Edsel] and thought he was improving," explains Cousins. "The next morning, as we drove through the gate (we were four in a car, sharing rides in those war days), we saw the flag at half-staff. We knew it was Edsel. We had lost our man. We all started crying."

While driving to work later that same morning, Gregorie heard on the radio that Edsel had died. Similar to everyone else, Gregorie was caught completely off guard by the announcement. Edsel had done a good job at keeping his serious malady a secret, even from Gregorie.

When Gregorie arrived at the design department, Joe Galamb met him. Somberly, Galamb told Gregorie, "Vee loost our boss."

"Yes, I heard," replied Gregorie. "I guess we'll just have to see what happens now."

"The old man [Henry] was always an anchor around Edsel's neck, as far as the product was concerned," Gregorie solemnly recalls. "The old man didn't know a thing about design, but he was an obstruction in the way of design, and he had to be reckoned with. And that was an unfortunate situation. I think that is what ultimately killed Edsel—worrying about how to handle the old man. But the old bastard held on to the end. Oh, well, all you can do is dream about it."

Postwar Development

Figure 17-30. This is the last photograph of Edsel Ford, taken on March 12, 1943, only a couple months before Edsel's untimely death. Here Edsel is being presented the Industry Award from Colonel Alonzo Drake for the contribution of Ford Motor Company to the war effort. Gaunt from his battle with cancer, Edsel appears much older than his 49 years should indicate. Soon after this photograph was taken, Edsel went to see Gregorie. They talked for only an hour, and then Edsel left. Gregorie never saw him again. (Photo courtesy of The Detroit News, © *The Detroit News)*

Edsel never lost his passion for automobiles, and he maintained a stable of motorcars until the time of his death. It is unknown how many motorcars Edsel had owned during his lifetime—probably thousands—but we do know how many he had at the time of his death. Parked in the garage of his Gaukler Pointe estate were three 1941 Mercurys (a station wagon, a coupe, and a sedan), two 1941 Fords (a coupe and a sedan), a 1941 Lincoln Continental cabriolet, and the second continental car that Gregorie had designed for Edsel almost a decade earlier.

A few weeks after Edsel's death, Henry Ford unexpectedly walked into Gregorie's office early one morning. Henry rarely came to the design department when Edsel was alive. Now that Edsel was gone, Gregorie thought it curious that the old man should appear, and he became suspicious.

"Let's take a ride," said the motor magnate. "I want to discuss something with you."

Gregorie did not know why Henry wanted to go for a ride with him, but they went to the service garage behind the engineering laboratory and climbed into a new Ford sedan.

"Where would you like to go?" asked Gregorie.

"Let's go down to the test track," directed Henry.

As they drove up to the entrance of the track, Gregorie slowed down to check in at the gate, but Henry instructed Gregorie to drive right through the gate. "They're all asleep anyhow," Henry growled. Then, pointing to the left, Henry said, "Come on, let's go on around there."

"But, Mr. Ford!" yelled Gregorie, "that's the wrong direction!"

"That's all right. That's all right," Henry replied defiantly.

Gregorie steered the Ford into oncoming traffic, and test cars immediately began swerving and skidding to avoid hitting the errant Ford coming toward them. Several cars almost collided with Gregorie's car. "Fortunately," says Gregorie in relief, "one of the gatemen saw Mr. Ford in the car with me, and he passed the word very quickly what was going on. Danger lights began flashing, and all the traffic came to a standstill. We could have been the victims of a horrible thing. We could have been killed if one of those test cars had hit us. It came that close."

As Gregorie drove back to the engineering lab, old Henry acted as if nothing had happened and tried making small talk with Gregorie, asking about current events, how the car was running, and such things. Gregorie was cautious about saying too much, for fear that he might mention something he would regret. "I wasn't quite sure what the old gentleman wanted, because he had never talked to me before," explains Gregorie. "I just told Mr. Ford that everything was going smoothly in the design department, and that we'd appreciate him stopping by whenever he felt like it—we'd show him what was going on.

"In all the years that I worked there," continues Gregorie, "that was the first and only meeting I had ever had with Henry Ford, and it came directly from him. In other words, I never prompted it. I never made any attempt to get in contact with Henry Ford—to offer my condolences to him—after Edsel had died. I thought the more I laid low, the better off I'd be. So I let him do all the talking. But I had a feeling that he wanted to talk to me about Edsel. He knew that Edsel would come to see me after he had lunch with him and the other Ford executives, and I think he wanted to find out how much influence I had had over Edsel. He always thought that I was the one who put newfangled ideas into Edsel's head about design and new models and so on.

"I assumed that the old man figured that I might want to have a personal discussion about Edsel and waited for me to bring him up. But I didn't bring up Edsel. If he wanted to bring up Edsel, I preferred that it came from him first. But he didn't talk about Edsel, either. For some unknown reason, he clammed up at the last minute, which suited me just as well. I didn't want to get into a discussion pertaining to Edsel."

When Gregorie pulled into the garage, Henry Ford got out of the car and walked away. "It was a peculiar instance," continues Gregorie. "It left me with a strange feeling. I didn't know what was going through his mind. (Neither did a lot of other people!) I didn't see the old man much after that.

"Oh, the old man was crazy!" adds Gregorie. "He was still in a daze over Edsel's death.... He had put Edsel through the ringer for years. Edsel would come into my office after lunch and just shake his head in frustration. He never told me any details of the discussions that took place during lunch with his father, but he didn't have to. I could tell that they weren't pleasant just by the expression on his face and the tone in his voice."

With Edsel's death and with Henry Ford all but out of the picture because of his age and ill health, a tremendous vacuum developed within Ford management. "You see," explains Gregorie, "there was a great big mad scramble for the spoils of the company, and the executives were at each other's throats. The Lincoln people were jealous of the Ford people, and the Ford people were jealous of the Lincoln people, and Harry Bennett was there trying to get hold of the big grab bag at stake. The whole official structure of the company was turned topsy-turvy. Things were in complete turmoil."

Two years earlier, Henry Ford II, then only twenty-three and recently married, enlisted in the Navy reserve in case the United States entered the war in Europe. Three months after Edsel's death, the young Ford was released from the Navy by Secretary of War William Knox to help take control of the company. However, young Henry had been assessing the situation in the company since his father's death.

Henry II had never seemed interested in the automobile business, but apparently Edsel had tutored him more than anyone had realized. In particular, Edsel had explained to Henry II the importance of automobile design in the continued success of Ford Motor Company. The first thing young Henry II did when he came to evaluate the situation was to visit Gregorie. He walked into the chief designer's office, shook hands, and said, "Father told me to start here."

"That's amazing, when you think about it," says Gregorie. "Here Edsel was on his deathbed, and one of the things he wanted to impress upon his son was the importance of the design department to the future of the company. It shows just how much he loved the design of automobiles. It also says a lot about what Edsel thought about the working relationship we had developed over the years. I was moved by the thought."

Henry II and Gregorie had quite a discussion, with Gregorie bringing the new Ford boss up to date on his postwar plans. They also drove to the Lincoln plant, where they met with the manager, Robby Robinson, who explained to Henry the problems he was having in getting the tooling built for the new Lincolns. With Henry II showing such interest, Gregorie hoped and believed that perhaps the young man would pick up where his father had left off. Gregorie had known Henry II since the new boss was a little boy. Henry II and his younger brother Benson had come to the design department with Edsel on many occasions, and they would roller skate among the clay models. With such an intimate relationship, Gregorie and Henry II should have been able to build a collaboration similar to the one Gregorie and Edsel had developed over the years.

Unfortunately, after that initial meeting, Henry II never showed much interest in car design. "He wasn't even conversant with it," says Gregorie. "He acted bored with it. From the day he started there, even though his father told him to start in my department, I really thought he had a better grasp than he had. He was just more or less disinterested in the design of the product. You couldn't interest him in a new design or discuss the design of a new car with him. He was more impressed with the big issues of the whole company. Design was just part of the big problems. I could never get his attention long enough to discuss design with him." Whenever Gregorie tried to get Henry II involved in a design decision, Henry II would say nonchalantly, "Yeah, go ahead with it," without really understanding or caring about what he had approved. "That was too bad," laments Gregorie.

At that point, Henry II was still in the Navy and would not be released for another two months; therefore, he was not in a position to protect Gregorie from the designer's old nemeses. Old Henry Ford's men, especially Joe Galamb, had never forgotten how Gregorie, with Edsel's support, had eroded their power over the years. Galamb was the one who had originally worked with Edsel on the design of new cars, and his input into their design had been all but eliminated when Edsel appointed Gregorie design chief. Galamb had always resented Gregorie for that. However, as long as Gregorie had Edsel to protect him, Galamb stayed clear. Now that Edsel was gone, Galamb returned with a fury.

"Joe gave me a bad time after Edsel died," Gregorie says. "He thought, 'Oh, boy, I'm really going to fix this guy!' " Galamb started coming in almost every day, telling Gregorie what to do, which was precisely what Gregorie had feared. He and Edsel had fought hard to separate design from engineering, and now they were back at his doorstep. "I didn't want to get involved with it," says Gregorie. "That was one thing I fought with Sheldrick and the other engineers about when I was a kid. I'd finally gotten design split apart from them. I put design on more or less of a separate department under Edsel and myself. And, man, I wasn't about to step into *that* again." Gregorie did all he could to keep Galamb at bay, but it eventually became a lost cause. Galamb wanted revenge. He wanted Gregorie out of Ford!

Until this time, Gregorie had held his own against the Napoleonic Galamb. Occasionally, the defiant design chief even taunted Galamb. For example, every fall, all the top brass of the automobile companies attended the New York Automobile Show, where they first introduced their latest automobile designs to the public. It was the event of the season, and the New York Central railroad had to run two *Detroiters* simultaneously end to end from Detroit to New York to accommodate the added traffic. When the two streamliners were approximately twenty miles outside New York City, the steam engines were exchanged for electric engines for the ride under Manhattan, and the two trains ran side by side on parallel tracks into Grand Central Station. "Frequently, you'd be sitting there having a drink," recalls Gregorie, "and you'd see some guy that you knew in the next train, and you'd start making little signs back and forth to each other—it became quite an event. Anyway, I was in the parlor car having a drink during one of those trips, and I looked over and saw Joe Galamb sitting in the parlor car of the other train. He didn't like the idea of me going to New York. He always frowned on me, ever since I started working for Edsel. We made eye contact, and I lifted my glass and nodded my head in a toast. Boy, did that irk him! He just sneered and looked the other way."

Unfortunately, now that Edsel was gone, Gregorie had lost the only man in the company who could protect him. The end came one Friday afternoon in June 1943, while Henry Ford II was still in the Navy. In the months following Edsel's death, Galamb not only began taunting Gregorie, but he also began looking for someone on the company board of directors to fire him. Galamb found his ally in Frank Campsall, old Henry's confidant. Campsall called Gregorie to his office in the Ford administration building and fired him.

Gregorie was not too surprised. In those days, Gregorie says, "firing of officials was popular—it was a trend. They got rid of Sheldrick and Sorensen at about that same time. I got notice from Frank Campsall—that's all I know about it. I don't recall ever having any contact with Henry Ford II, or I think he would have intervened."

During this same period, Gregorie had no one to direct him or to approve his designs. "So I was left at a standstill, and it was awkward for me," he explains. "There were no other company officials that would assume any responsibility. The place hadn't been reorganized on any basis—they hadn't any committees or anything like that. That came along later. But I'm afraid it came along in such a way that it wouldn't appeal to me at all. I liked the simplicity of working with Edsel Ford. But now, it was an entirely different situation." From his meetings with Henry II, Gregorie realized that he soon would be forced back under the direction of engineering by the sheer vacuum created by Edsel's death. "I accepted my fate as it came," says Gregorie. "I couldn't get anything done anyway. It was all right with me."

Instead of returning to New York or taking an extended vacation as he did when he was laid off in 1933, Gregorie set up his own design firm in a penthouse suite on the twenty-eighth floor of Eaton Tower in downtown Detroit. "I rented these rooms, which had a beautiful view," says Gregorie. "You could see over to Canada and the ships in the river." He won contracts with Rohm & Haas, a plastics supplier, to design and develop plastic automotive body trim, and with Nash-Kelvinator to design appliances. "I worked a couple of hours a day, drawing radio knobs or ice boxes," he says. "It paid the expenses." Gregorie must have had ample work, for he paid John Walter, the man who designed instrument panels and steering wheels at Ford, to help him on his own time.

It appeared as if Gregorie was going to make a smooth transition from corporate designer to independent consultant, in the same way that Harley Earl and Bill Mitchell from General Motors would do later. However, everything changed when Gregorie received a phone call from Henry Ford II a year or so later.

CHAPTER EIGHTEEN

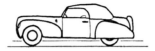

THE 1949 MERCURY AND LINCOLN
The Last Hurrah

When Henry Ford II was released from his Navy obligations in August 1943 and began working at Ford on a full-time basis, he found that Gregorie had been fired. "Tom Hibbard came up in the vacuum that was left when I left," explains Gregorie. "He was down at the Rouge plant, and Joe Galamb brought him up to do my job. But I don't think he came over right away. I didn't designate anyone in particular when I left. It was kind of a dead period, anyway."

For four months, Henry Ford II had worked in the company with no official title and no particular duties. Throughout most of that time, he simply wandered through the Rouge plant and asked questions. However, in December, he was appointed vice president by his grandfather, and, from that moment, the young Ford quickly assumed responsibility. Although young and inexperienced, Henry II learned the business quickly. From the few experienced and loyal managers that had survived his grandfather's purges, Henry II established a policy committee to help him run the beleaguered company. This was only an interim step because Henry II wanted to weed out the old regime and recruit a group of young, college-educated men to become his executives. Until that happened, he would run Ford Motor Company through this new arrangement.

Henry II and the policy committee had to address many problems, and months passed before they began to evaluate the design department. Henry II was well aware of the product plans that his father and Gregorie had devised a few years earlier, but John Davis was the only member of the original team who was still there to implement the small Ford/big Ford program. Davis could sell the program to the dealers and to the public, but he could do little to complete the design of the cars that Edsel and Gregorie had conceptualized. Clearly, Henry Ford II needed Gregorie.

"Henry Ford II called me, and he wanted me to come up to his office," recalls Gregorie. "He said, 'Now, Bob, we'll get things going again.'" At first, Gregorie was reluctant to meet with Henry II, but he could not refuse a personal invitation from Edsel's son. Gregorie went to see Henry II the next morning.

During their meeting, Gregorie explained to Henry II the obstruction he began receiving from engineering as soon as Edsel died. He described how it took him and Edsel years to insulate their work from the petty and unnecessary interference of engineering in the past, and he stated that he would not care to work under those old conditions again. (Henry was receptive because he knew that at General Motors, design worked independently from engineering.) Gregorie said, "I'll come back on the condition that I won't have any interference from certain people that I know will create problems for me and the department."

"Like who?" asked the young Ford.

"Well, Joe Galamb, for one," answered Gregorie.

"I'll get rid of Joe right now!" exclaimed Ford.

"We don't have to do it that way, Mr. Ford," Gregorie replied, feeling a little sorry for Joe now. "Can't you give him something else to do, or tell him to stay out of my area?"

"Naw," replied Ford, "I'll get rid of him."

"That's the way he was," says Gregorie of the new boss. "He was impulsive as hell. He grabbed the phone, called somebody, and said, 'Get rid of Joe Galamb. Get rid of him. I don't care what you do. Get rid of him!'"

It was a sign of changing times within the company. Joe Galamb had been a close associate of old Henry since 1905 and had helped develop the Model N and the Model T. Now, he was out of the company, fired by Henry Ford's grandson. However, Galamb did not have to worry about money. Ford had paid him well over the years, and rumors claimed that he was the richest Hungarian in Detroit.

With a salary twice what he had been making prior to his termination and with engineering cleared from his way (or so he thought), Gregorie closed his design firm and returned to Ford in April 1944, after an absence of ten months, regaining his position as chief designer. (Figure 18-1) Tom Hibbard was a little surprised when he was told that Gregorie was returning to take over the department again, but he did not raise any argument. When Gregorie entered, the two men introduced themselves to each other, and Hibbard asked, "What would you like me to do?" Gregorie was well aware of Hibbard's background in the custom-body business; therefore, instead of returning Hibbard to the Rouge plant or relegating him to a drafting table, Gregorie made him his assistant.

Most of the men who were in the department when Gregorie was fired were still there when he returned, and the entire team continued from where they had left off. Because the little Ford and big Ford sedans were already in the tooling stage, Gregorie and his men began a flurry of activity to complete the Mercury and Lincoln lines they had started more than a year earlier. By June 1945, full-size clay models of Gregorie's new Ford, Mercury, and Lincoln lines were ready to be shown to company officials and plant managers throughout the country. All these cars should have been complete working prototypes by this time, but the engineering department had delayed in developing new suspension systems and chassis for the new models. For example, when the first prototype of the big Ford was completed, the old buggy suspension was still under it. "We were way ahead of engineering," says Gregorie. "We had the appearance of the new cars for three years before we got new chassis to go along with them. Nevertheless, we had to keep sufficiently flexible to accommodate any areas that were questionable, such as the floor pan, the tunnel for the drive shaft, and other heavy mechanical aspects of these cars. As it turned out, the engineering for these cars was not started until the summer of 1946."

Henry II was glad to have Gregorie in the company again, but he had concerns other than design about which to worry. By the time World War II ended, Ford Motor Company was losing $10 million a month, and Henry II realized that he needed expert help to stop the bleeding and save the company. Many of the

Figure 18-1. Shown here are members of the Ford Motor Company engineering board, as of November 1945 (left to right): Dale Roeder, Ford chief engineer; E.T. Gregorie, styling director; Russell McCarroll, executive engineer; Val Tallberg, administrative engineer; Clyde Paton, Mercury chief engineer; and Jack Wharam, Lincoln chief engineer. Gone from the scene were Henry Ford's cronies and Gregorie's nemeses—Larry Sheldrick, Charley Sorensen, and Joe Galamb. (Photo courtesy of Ford Motor Company)

their "car for every purse and purpose" strategy—but also its highly successful management techniques. Edsel must have emphasized this to his son over the years, because Henry II also thought highly of how General Motors was run. If he could recruit some General Motors executives to Ford, he thought, they could help him reorganize Ford Motor Company into the image of General Motors.

After months of negotiations, Henry finally coaxed one of the brightest executives at General Motors to help him run Ford. Ernest Breech, a self-taught accountant from Arkansas, had risen in only two years to the rank of general manager of the General Motors subsidiary, Bendix Aviation Corporation. Bendix had been losing $3 million a year when Breech took control. Two years later, it was making an annual profit of $5 million. By the end of the following year, 1945, Bendix posted a net profit exceeding $55 million. It was commonly believed that Breech's next move would be to head Chevrolet and eventually become president of General Motors. From a managerial standpoint, Breech was a logical choice for Henry Ford II.

old Ford executives were nearing retirement, and Henry II was not confident that their subordinates had the wherewithal to undertake the monumental tasks that confronted the company. Therefore, he looked outside the company, particularly at General Motors, for new management.

For years, Edsel Ford had tried to emulate the great General Motors, not only from a marketing standpoint—trying to copy

Partially through Henry's persuasive manner and partially through the challenge that Ford Motor Company offered him, Breech quit General Motors and joined Ford in July 1946. He brought with him three other men who were experts in finance, engineering, and manufacturing. They, too, had all achieved high positions within General Motors and were willing to tackle the Ford challenge that Breech offered to them.

First, Lewis Crusoe was a highly talented finance man who had been assistant controller of General Motors. He was Breech's assistant at Bendix when he was asked to join Ford. Second, Harold T. Youngren had been an engineer at Oldsmobile and was instrumental in developing the Hydra-Matic transmission. He was chief engineer of Borg-Warner when he received the call from Breech. Third, Delmar Harder was a manufacturing expert. He had served with Crusoe for years at Fisher Body, and when Breech asked for his help, he was president of the E.W. Bliss Company.

Each of these men immediately attained high positions in Ford when they joined Breech. Crusoe became the chief financial officer at Ford, reporting directly to Breech. Youngren became the new chief engineer at Ford, replacing Larry Sheldrick who had been fired around the time Gregorie had been fired. (Sheldrick went to work for the General Motors Detroit Diesel Division.) Harder became vice president of operations at Ford.

As for Breech, Henry Ford II made him executive vice president and gave him a seat on the board of directors. From that point onward, Ernest Breech called the shots at Ford Motor Company and began making swift and monumental changes. "A new broom sweeps clean," is the way Gregorie puts it.

At first, Gregorie welcomed the changes. As far as he was concerned, before Breech and his men joined Ford, conditions within the company were not much better than what they had been after Edsel died. Although Gregorie no longer had to contend with petty obstructions from engineering, some semblance of an organization had existed in the old days. When Gregorie returned to the company in 1944, "the whole place was a nuthouse," he says. It was not much better in 1946, before Breech joined the company. During meetings of the policy committee, of which Gregorie was a member, he could see that most of the people on the management team were not worth their salt. Gregorie had his hands full in the design department and could not do much to assist Henry II; however, he did advise young Henry to start looking for new talent to help run the company. "I put a bug in young Ford's ear one Saturday morning," Gregorie recalls. "I told him we were badly in need of a top management rearrangement, that we needed some new department heads, and we needed some new thinking in the company itself. And I think, as a result of my prodding, he went out and got Breech. That's a fact! I told him things were in a bad state, and he couldn't be expected to come up with the ideas to reestablish the company himself. He was just a kid that came in there. He had to listen to everybody's story, and he was so dammed confused he didn't know whether he was coming or going. He needed a man just like Breech. I didn't like [Breech]—he was a little man, a Napoleon-type, assertive, you know—but he got things done. He knew how to stimulate things. Even though they didn't go my way, I have to give him credit for getting the wheels going. He was a good businessman."

One of Breech's first tasks was to review the Ford product lineup for the postwar market, and he was not impressed. He could not see the need for the small Ford, which he thought was too small, and he felt the big Ford was too big. Youngren had told Henry II and Breech that General Motors was making every effort to minimize weight, and thus that should be the focus at Ford also. In addition, Youngren suggested to the two executives that "we should not place all our hopes on Mr. Gregorie." Breech suggested that the company hire an outside styling consultant to review Gregorie's designs.

George W. Walker was an industrial designer whose primary experience in the automobile industry was designing hardware: door handles, window cranks, and such items. His greatest assets were his gift of gab and his friendship with Ernest Breech.

Breech himself drove to Walker's office in Detroit and apprised him of the situation. "We want you to come out to Ford and take a look at our products and give us your opinion on them," Breech said.

Walker went to the Ford engineering lab, and he met with Breech and Henry Ford II. The three of them walked around the various prototypes and clay models on display, and eventually Walker brought them back to Gregorie's big Ford. "That's terrible," Walker said. "You'll go broke doing that."

"Well, we're going to put that one out," said Henry II.

"Don't do that," replied Walker. "It's a heavy thing. It's all out of shape. It looks like a pregnant Buick." (Figures 18-2 through 18-6)

"Could you do any better?" Breech asked.

I could do better with my eyes closed," replied the cocky designer. He wanted the account desperately, but he was already under

Figure 18-2. This is an early quarter-size clay model of a postwar Mercury. It has an expansive greenhouse area, which is what Gregorie was seeking, but the entire design was soon abandoned for a less bulky one. (Photo from the Collections of Henry Ford Museum & Greenfield Village)

Figure 18-3. This clay model of the "big Ford" coupe is under development. (Photo from the Collections of Henry Ford Museum & Greenfield Village)

Figure 18-4. Here is the full-size clay model of the "big Ford" coupe. (Photo from the Collections of Henry Ford Museum & Greenfield Village)

Figure 18-5. This photograph shows the full-size clay model of the postwar Custom Mercury Tudor sedan, circa June 1945. "Both the Custom Mercury and the Lincoln were to have pop-up headlights," explains Gregorie, "but the engineers couldn't get a mechanism to work well enough by the time the cars were ready to go into production. Therefore, they filled in the hole where the mechanism was to go with a strange-looking chrome insert." The Custom Mercury was never produced. (Photo from the Collections of Henry Ford Museum & Greenfield Village)

Figure 18-6. This is a pre-production prototype of a postwar "big Ford" (later as the 1949 Mercury) made from production tooling. This photo was taken in November 1945 and confirms Gregorie's statement that the tooling for the little Ford and the big Ford were almost finished by the time Ernest Breech was hired by Henry Ford II in July 1946. (Photo courtesy of Ford Motor Company)

contract with Nash-Kelvinator and International Harvester, and each was paying him $75,000 a year.

"Well, think it over," requested Breech.

A few nights later, Walker met with Breech at the Bloomfield Hills Country Club.

"Well, have you made up your mind?" Breech asked. "It's a great opportunity. I can't tell you what's going to happen, but you can make yourself millions of dollars if you come with us."

Walker was ready for the offer. He had already spoken with both Fowler McCormick at International Harvester and George Mason at Nash-Kelvinator, and both agreed to let him out of his contracts. When Breech asked again, Walker replied, "Well, fine, I'll do it."

By the time Walker entered the picture, the tooling was almost complete on the little and big Fords. If Walker had had his way, he would have scrapped both cars and started from scratch. However, the company had already spent millions of dollars tooling these two models, and it was running out of time to develop a new line of cars for the postwar market. It needed to find a way to save some of Gregorie's work.

Then Breech had an idea. At the moment, he didn't know what to do with the little Ford, but he suggested to Henry Ford II and to Walker that they should make the big Ford the new Mercury and start afresh on a new Ford design.

"Yes, that's what you should do," concurred Walker.

However, if the little Ford were too small and the big Ford too big, what should be the size of the new postwar Ford? Youngren settled that question as old Henry Ford and his chassis people used to settle it for Gregorie in the old days. However, Youngren set many more parameters for the car, not only the length and width of the chassis, as Henry Ford had done. "Youngren and his boys bought two 1947 Studebakers and dissected them from every angle, and they arrived at what they felt was the proper hip room and head room, length, width, and height," explains Gregorie. "So, what we were given to work with was that formula, or package, all based on that little, double-ended Studebaker." (Figure 18-7)

At Breech's directive, two design groups were set up—a sort of design competition—one headed by Gregorie and the other by Walker. Each group would develop a new Ford design based on Youngren's dimensions. The policy committee, headed by Breech, would determine which of the new designs would become the new postwar Ford.

Gregorie had the resources of his design department at his disposal, but he worked primarily with Tom Hibbard to develop a design. Walker, working from his office in the New Center Building across the street from the General Motors headquarters in Detroit, delegated the design efforts to three talented designers on his staff: Joe Oros, Elwood Engel, and Holden Koto. As usual, Gregorie did much of the design work himself, drawing on Hibbard for ideas and feedback. On the other hand, Walker never lifted a pencil in developing his design.

Meanwhile, the policy committee had agreed that design and engineering should make an all-out effort to produce Gregorie's small Ford. Nonetheless, Youngren balked. He explained to Breech that General Motors had abandoned work on a small car, primarily because the company felt the market was too small to pursue. Youngren was supported by a recent survey conducted by one of Breech's own men. What Elmo Roper had found was that, in this postwar era, most Americans wanted larger cars and were willing to pay for them. With this new information, the policy committee retracted its edict of the preceding month and decided to abandon the small car project for the American market. To save the tooling of the small Ford, it was shipped to France, and Gregorie's small American Ford became the 1948 Vedette from Ford of France. (Figures 18-8 and 18-9) In fact, the design was yanked from Gregorie

Figure 18-7. This 1947 Studebaker—the first all new postwar model in the industry—was purchased by Ford Motor Company in the summer of 1946 for engineering analysis. Shortly thereafter, Youngren ordered Gregorie to design a Ford around similar dimensions. (Photo from the Collections of Henry Ford Museum & Greenfield Village)

before he and his design staff had a chance to complete it. "We didn't do any of the final detail work on that small Ford," explains Gregorie. "The basic tooling was done, but details like the bumpers and applied trim were done in France."

By now, both design teams had completed quarter-size models of their design proposals, and Gregorie and Walker were instructed to begin work on full-size models. Because Walker did not have the facilities or manpower to construct a full-size model, Breech allowed him to set up an area in the north end of the engineering building and to borrow some of Gregorie's clay modelers to construct his model. According to Walker, "There was supposed to be a guard there all morning, noon, and night to make sure nobody entered the area that we had set up. But Youngren allowed Gregorie and John Najjar to go in one evening and make sketches of what we had." His implication was that Gregorie was stealing his design.

Gregorie recalls this time much differently, and he adamantly insists that neither he nor his designers were banned from Walker's area. Likewise, Walker's men were not forbidden from entering Gregorie's work area. In fact, no secrecy whatsoever existed between the two work groups. From the outset, Gregorie knew about Walker's work, and Walker knew about Gregorie's work. "We were a party to it," explains Gregorie. "My men set up the partitions! My men did the modeling! It was the damnedest thing the way it worked out."

Figure 18-8. Despite this full-size working prototype of Gregorie's "little" Ford design, this car was never made in North America. The tooling was sent to France, and Ford of France placed this car on the market as the 1948 Vedette. This photo was taken in December 1944. (Photo from the Collections of Henry Ford Museum & Greenfield Village)

By November 22, 1946, both groups had clay models ready, and on December 11, Henry II, Breech, Youngren, and other members of the policy committee examined the two models. After a brief discussion, the members chose Walker's design over Gregorie's. Nobody was supposed to know who designed which car—the choice was to be totally objective. However, Gregorie knew this was not the case. "Breech ran the policy committee," Gregorie exclaims. "They were all beholden to him. They're not going to say, 'Gee, let's take Gregorie's car.' No, that would have been impossible. I realized that."

Figure 18-9. This version of a pre-production prototype of Gregorie's "little" Ford has a horizontal grille, whereas another version sported a rolled grille. (Photo from the Collections of Henry Ford Museum & Greenfield Village)

So much similarity existed between the two cars that the policy committee did not have to be too concerned about choosing the wrong one. "I couldn't see a lot of difference," says Gregorie. "They were both designed to a formula: the hood was to be so long, and the body was to be so wide, and the tread was fixed on the wheel size. We came up with practically the same sheet metal design. You couldn't change it very much. The grille was different, and maybe the hubcap design was different, but the car—you put one alongside the other, I couldn't tell one from the other."

Gregorie is right. Both cars were of the "slab-sided" design, and both had a similar greenhouse. However, Gregorie's design was slightly more rounded than Walker's design, the grillwork was different, and more definition existed in the side panels. "There was a little more shape in my car," explains Gregorie. "It tucks under a little, and it had a sheer line running the entire length of the body to give it a little form, whereas Walker's design has a raw edge—the side panels were completely smooth. The grille on my car, of course, was subject to change. We had several versions ready to try but never got a chance to evaluate them on the model." (Figures 18-10 through 18-12)

What most distinguished the two cars from one another was their paint jobs. Gregorie's car was painted a conservative bluish green, a color carried over from the Edsel days, whereas Walker's car was painted a bright yellow.

Shortly after the policy committee approved Walker's design, Henry Ford II brought his grandfather to the design department to show him the 24th-generation Ford model. Old Henry was not too impressed. Reminiscent of the garage scene from the 1920s, the old man walked around the model a few times. It was too low, and he could not see how a couple of milk cans could fit into that small trunk. When he tried to open one of the doors, the mock handle fell off in his hand. "That doesn't work," he said. Young Henry laughed and then escorted the founder back to his 1942 Ford sedan to be chauffeured home.

Gregorie reflects on this tumultuous period with humor. "It's better than crying," he says. "I realized it was a hopeless situation. Breech wanted to put on a show for Henry Ford II and the board of directors. There was never any doubt that their version of the car would be selected. It was foolish to even try to compete with it. We didn't have enough leeway; we were both heading in the same direction. I stopped the thing halfway through; it was senseless to duplicate their efforts. I didn't make any effort to finish up the car we were working on; it was foolish. It was just a gesture on our part."

When the design of the basic Ford was settled, Gregorie and his staff completed the design of the big Ford (now the Mercury) and the big Lincolns. One of the first things they did was to revamp the front end of the Mercury. Initially, all Gregorie's postwar models had conventional grilles with horizontal bars similar to the 1946 Fords. However, the sheet metal on these new models had a tucked-in appearance in the grille opening from the outset. At the last minute, Gregorie eliminated the horizontal grille bars and replaced them with a long horizontal grille piece made with a multitude of fine vertical convex-shaped grille bars. This design mimicked the rolled sheet-metal opening and "let the grille sort of float in there," describes the designer. Gregorie had made a similar grille on one of his Lincoln clay models, and it looked so good there that he decided to adapt it to the new Mercury also.

The postwar Lincoln designs went through a similar metamorphosis at the end. (Figures 18-13 through 18-21) Gregorie started with massive horizontal grilles that appeared almost solid, but he decided on a more conventional design that had two large horizontal bars. The most dramatic change occurred with the headlights. The original intent was to have hidden headlights on the Lincolns—a design cue not new to the industry but unique in the postwar market—so Gregorie designed the front end to accommodate such a feature. "The Lincoln engineers were to work up the pop-up mechanism," explains Gregorie, "but they weren't successful at getting the covers to open right. So at the

Figure 18-10. These photographs compare the left profile views of Walker's 1949 Ford design (top) to Gregorie's 1949 Ford design (bottom). The basic shape of both cars is almost identical, and both can termed a "slab-sided" design. However, Gregorie's design has an indentation along the entire length of the side "to give it form." (Photo from the Collections of Henry Ford Museum & Greenfield Village)

The 1949 Mercury and Lincoln

Figure 18-11. Front view of Walker's 1949 Ford design (top) compared to Gregorie's 1949 Ford design (bottom). Both designs had very little trim, and both sported a horizontal grille. However, Gregorie tried to add more form to the hood of his design by carving a "speed line" down the center. (Photos from the Collections of Henry Ford Museum & Greenfield Village)

Figure 18-12. Rear view of Walker's 1949 Ford design (top) compared to Gregorie's 1949 Ford design (bottom). Again, the similarity of the two designs is apparent in this view: the greenhouse is similar, the side glass is similar, and the back glass is similar. However, Gregorie's design has more form to it, and the speed line that he started on the hood is carried back to the trunk lid. (Photos from the Collections of Henry Ford Museum & Greenfield Village)

Figure 18-13. This is a full-sized clay model of Gregorie's postwar trunk-back Lincoln design. "This body is pretty well established," explains Gregorie, "except for the pontoons over the wheel wells. We considered both body-colored pontoons and chrome-plated pontoons, and the production model had chrome pontoons applied in these areas; there were no bulges in the sheet metal as shown in this model." (Photo from the Collections of Henry Ford Museum & Greenfield Village)

last minute, in a desperate attempt to get the cars into production, they designed a cast metal insert to fit into the headlight holes as a substitute for the hidden headlamps. That's why the Lincolns look a little strange in that area with the headlights recessed into the sheet metal."

Along with their new forms, the new Ford models all had new chassis. Gone at last was Henry's favorite buggy suspension, which he had used on his cars for half a century. New coil-spring, independent front suspension on the front and semi-elliptic springs on the back finally replaced Henry's favorite transverse-spring and wishbone, buggy-type suspension.

Shortly before Gregorie threw the Mercury and Lincoln designs over the wall to engineering, he decided to leave the company. "I couldn't work with Walker there," he says. "Of course, Breech had sold Henry II on the idea that there should be a whole new set up, and my connection with Henry II had dulled with Walker

Figure 18-14. Here is a full-size clay model of Gregorie's postwar fastback Lincoln design. Edsel liked fastback designs, and a model similar to this went into production as the 1949 Cosmopolitan town sedan. (Photo from the Collections of Henry Ford Museum & Greenfield Village)

Figure 18-15. Clay modeler Arthur Karpeles gives the final touches to a hood ornament on the proposed postwar Lincoln, circa 1945. (Photo courtesy of Ford Motor Company)

Figure 18-16. Clay modelers Clyde Trombley (left) and Werner Framke (right) take offsets from a full-size clay model. (Photo courtesy of Ford Motor Company)

Figure 18-17. Here, designer/illustrator Vern Breitmayer works on a full-size rendering of a postwar Mercury grille. Note the variations of Mercury hood ornament ideas. (Photo courtesy of Ford Motor Company)

Figure 18-18. Clay modeler Al Kellum works on a quarter-size clay model of a postwar Ford. Hired by Gregorie after World War II, Kellum was a latecomer to the design department. (Photo courtesy of Ford Motor Company)

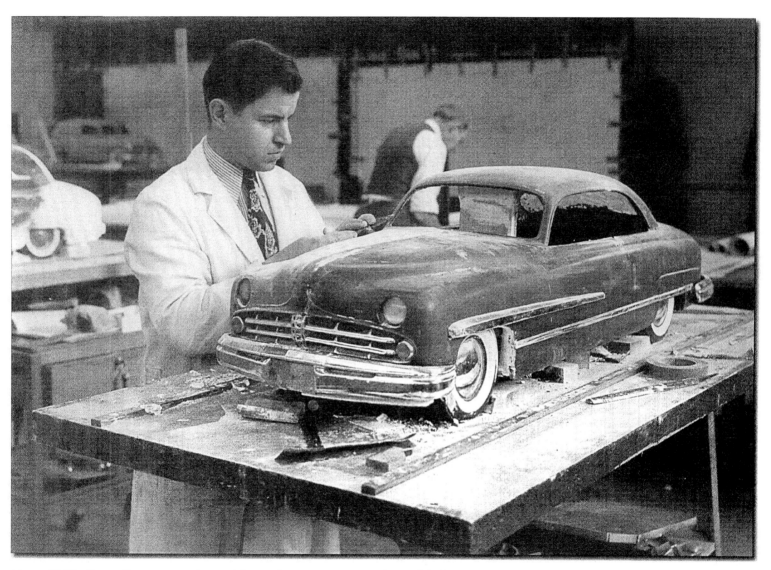

Figure 18-19. Cesare Testaguzza, a clay modeler, works on a quarter-size clay model of a postwar Lincoln. His brother Gino also worked in the design department. In the far background at the left can be seen a quarter-size model of a postwar sedan delivery. A standard offering of Ford's commercial line for years, the sedan delivery never became part Ford's postwar models, probably because of cost constraints and time limitations. (Photo courtesy of Ford Motor Company)

Figure 18-20. Designer Jim Huggins puts the final touches on a futuristic Ford design in 1945. (Photo courtesy of Ford Motor Company)

coming in. It was useless. The only thing for me to do was to make up my mind to get out of there. That's when I decided to go to Florida."

To Gregorie's credit, he decided not to make an untenable situation worse by battling with Walker. "I knew damn well that I wouldn't stand a chance if it came to a knock-down-drag-out contest between the two of us," Gregorie says. "So rather than be a part of it, I just decided to leave on friendly terms. I said to myself, 'If that's the way they want it, why, it's okay with me.' I looked forward to being relieved of it."

Gregorie had analyzed the situation correctly. If he had any doubt in his mind as to whether he had made the right move, Breech's remarks at a press conference introducing the new 1949 Ford set him straight. "This car," declared Breech, "was styled by George Walker and his staff, and don't you forget it. I don't want anybody to forget about that."

Gregorie is nostalgic regarding the big Ford that went into production as the Mercury. "The '49 Mercury was a very nice-looking car," Gregorie says. "It was one of the nicer designs we did. In fact, the big Ford that I designed for '49—which later became the Mercury—was only a little bit larger than the Ford that Walker and his men eventually designed. There was really no need to change my design. It was simply Breech's excuse to bring in his own men to design the postwar Ford and to gain power in the design activities in the company. Walker said that my big Ford looked too heavy, but he missed the whole point. If you can build a car that looks heavy and, at the same time, make it for a competitive price, why, you got something good to sell. You couldn't lose on it. Walker and Breech didn't comprehend that. The '49 Ford that Walker's group did will never be a classic. But the '49 Mercury—the big Ford that I designed—is already a highly sought-after car, and that puts a smile on my face. But what is more rewarding, looking back on it now, is that Edsel Ford had approved the clay model of this car. It was the last design that he and I did together. And the years have proven its timelessness."

As for the little Ford/big Ford concept that Gregorie and Edsel had conceived back in the early 1940s—the lineup that did not

Figure 18-21. On a cold, windy day in the fall of 1946, E.T. Gregorie (third from right) shows off his new 1949 Lincoln Cosmopolitans to Ford plant managers. The new models were displayed on the runway at the Ford airport, located down the street from the engineering lab and Gregorie's design department. (Photo courtesy of E.T. Gregorie)

impress Breech—that same approach came to fruition more than a decade later, long after Gregorie had left the company. In 1960, Ford introduced the compact Falcon to thwart the increasing number of small imports selling in the country and to offer an inexpensive and fuel-efficient alternative against the big cars that the entire industry was building. If Breech had thought that Gregorie's and Edsel's small car appeared too small at the time, he must have been shocked at the sight of the Falcon because it was almost thirty-three inches shorter than the standard Ford sedan! However, the car sold well—in excess of 400,000 units the first year alone—and ultimately became one of the most successful models produced by the company.

Although he had been gone for years, Gregorie did not miss the significance or irony when he saw the little Falcon in dealer showrooms. "Edsel and I had that same plan twenty years earlier!" Gregorie proudly explains. "If it wasn't for that damn Breech coming in and scrapping our plans, why, we could have come out with a similar product lineup about fifteen years earlier and caused a great stir in the industry. I think the success of the Falcon proves that our thinking of coming out with a small Ford after the war was both revolutionary and accurate. But as soon as Breech walked in the door of Ford Motor Company, he began poisoning Henry Ford II on all the things that Henry Ford, Edsel, and the rest of Ford management had implemented over the years. He wanted to erase any memory of them altogether. Surprisingly, and unfortunately, Number Two Ford went along with him. Coming from General Motors, Breech's one purpose in life was to GM-ize Ford Motor Company, and he immediately gave the orders to hire ex-GM hacks and stuff the company with them. Eventually, he was able to throw a blanket over any and all policies that had been put in place by Henry and Edsel Ford. No matter if their plans, processes, or methods were logical and reasonable, Number Two Ford and his men scrapped them all, at Breech's recommendation, in order to implement their own, even if they were untried. I never understood why Henry Ford II went along with Breech to the extent that he did. I guess he just wanted to disassociate himself from the old days and do things his own way. I suppose he didn't want to go down in history as being accused of not having a backbone, like his father." (Figure 18-22)

Figure 18-22. In the fall of 1945, Henry Ford II decided to show the world that he was in control of a new Ford Motor Company by authorizing a major construction program. During the next seven years, a new world headquarters, engineering center, and styling studio would be built on the land that was once part of Henry Ford's estate. During preliminary discussions with the New York architectural firm of Voorhess, Walker, Foley & Smith hired for the project, Henry Ford II asked Gregorie to sit in on the meetings. Gregorie is shown here at the far left. The third man from the left is Henry Ford II. To his immediate left is Russell McCarroll, and the gentleman at the far right is the Ford sales manager, John Davis. (Photo courtesy of E.T. Gregorie)

CHAPTER NINETEEN

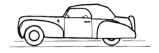

THE END OF AN ERA

"...An Extended Vacation"

Gregorie had known George Walker for years and held no personal grudge against him for winning the design contest. "It was Breech's sneaky tactic that bothered me," explains Gregorie. "Breech disliked my direct contact with Henry Ford II, just as I had worked with Edsel on design matters. Breech and Walker were golf buddies. They belonged to the same country club and were neighbors out in Birmingham. But I could see that Breech's tactic was to inject him in there as a consultant to me. In the meantime, Walker would be working for Breech. In other words, he wanted to drive a wedge between Henry Ford II and me. As a matter of fact, my activities were placed under Harold Youngren—new vice president of engineering. We got along fine. We saw eye to eye. In fact, we enjoyed seeing one another on our boats on the East Coast years later, and if Breech had only kept our two activities separate, why, it would have worked out just fine. But at the time, I found this new arrangement to be a very peculiar situation. I mean, it took me years to get design separated from engineering, and after getting rid of Sheldrick, who had been a monkey on my back for years, I now found myself working for an engineer! An engineer, mind you, not even a body designer! And a General Motors engineer at that! He had no design knowledge at all. The reason they brought him in here was because they were desperate for an automatic transmission. Youngren had been chief engineer at Oldsmobile and was instrumental in the development of the old Hydra-Matic transmission. That's why Breech latched onto him." Gregorie found the whole arrangement untenable. "I didn't want to get involved with it," he says.

Gregorie had had a short reprieve as long as he was reporting directly to Henry Ford II. But now that Breech was in control, with Walker acting as a spy and Youngren directing all of engineering and design, Gregorie lost heart. "I mean, I had such a happy relationship with Edsel Ford and working directly with him. We managed to arrive at some very workable decisions with a minimum of meetings and conferences and lost motion. You'd be surprised at the decisions that were made just sitting at the end of a drafting board. I was involved with a subject matter that he personally loved, and I enjoyed it as well. I began to feel a great responsibility, because Edsel depended on me to such an extent."

When Gregorie lost Edsel's backing, he lost the war. "I was the guy to get, see," Gregorie explains. "I was going into battle with a little tiny destroyer with which to sink the navy. Breech wanted to have a definite part in the design department. He wanted a spokesman that *he* could depend on. That's why he wanted Walker in there. Walker was a blustery, flashy type. He didn't know [anything] as far as design was concerned. I don't think he could draw a straight line. He never designed an automobile. He designed refrigerators, door handles, hardware, and stuff like that. I can see him sitting out in the lobby now, with some of that stuff to show Galamb.[a] Edsel Ford wouldn't have had any part of him. Edsel wasn't an exhibitionist, and he didn't like exhibitionist types around him. But that's the way the cards turned out."

Breech's treatment of Gregorie was a poor reward for a man who had been instrumental in saving the company from disaster during the Depression. We could reasonably conclude that with Henry Ford's antiquated engineering, Ford Motor Company might have succumbed to the Great Depression had it not been for the stylish designs created by Gregorie under Edsel's direction.

Four days after the policy committee approved Walker's design, Gregorie went to see Henry Ford II. "I told Mr. Ford, 'My goodness, there's no way this thing will work. You've given Mr. Breech full powers in all departments, and I know that he's not going to favor me—not when he's got his own men in here.' "

"It's pretty much out of my hands," concurred Henry. "We have Mr. Breech here, and he is making changes."

"I can well understand that," replied Gregorie. "I don't think it would quite fit my style. I think I would be an obstacle in the path of this changeover arrangement. And if it's all right with you, why, we'll part good company. Let's put it on a six-month basis. After six months, if I feel like coming back, I'll get in touch with you. And if you think you'd like to have me back...why, you see how things go. In the meantime, let this new crew make it work out. I don't mind giving up at this point. There are certain things I don't care for, but before we get into it and have a hassle over it, I'd rather not stand in the way."

Thus, on December 15, 1946, at the age of thirty-eight and just shy of his sixteenth anniversary with Ford Motor Company, E.T. Gregorie, the brash and talented boat designer turned automobile designer, returned to his two lost loves—boats and the sea. (Figure 19-1) "I sold most of my cars and everything else I didn't need," he says, "packed what was left, and headed for Florida. I wanted to get as far away from Detroit as I could. I decided to take an extended vacation." (His "vacation" has lasted for more than fifty years!)

Before leaving for the third and final time, Gregorie suggested to Henry Ford II that Tom Hibbard again be placed in charge of the design department. Youngren agreed, but as far as he was concerned, it would be an interim position until he could find someone more to his liking.

"Tom was a great gent," says Gregorie. "He had great experience, and he was very diplomatic in handling the employees. He had a good sense of humor and was highly efficient. I told him I couldn't take it any more. I hated to leave him in that embroilment, and I wished him the best of luck."

Hibbard kept in contact with Gregorie during the next few months and wrote to Gregorie shortly after taking over the design department, "It's hell here. It's awful here. Why don't you come back?"

"No, it's too late now," Gregorie responded.

Ironically, after Walker's design was approved, his contract with Ford was fulfilled, and he left. Although Breech said he would have other work for Walker to do, Walker had his own design

---

[a] In his Design Reminiscences at the Ford Archives, Walker recounts the first time he tried selling his automotive hardware to Henry Ford. He was there with a number of other vendors, and to make his highly polished door handles and other items stand out, he placed them on a black cloth. When Henry Ford walked in, he was immediately attracted to Walker's display.

Six months later, Youngren recruited John Oswald from General Motors to head the design department at Ford Motor Company. At that point, Hibbard quit.

Around this time, Breech coaxed Walker to return to Ford, again as a consultant, but this time to Oswald. In 1955, Walker became the first styling vice president at Ford. He retired from that position in 1961. Walker died at his Arizona home in 1993, at the age of ninety-six.

What prevented Gregorie from remaining at Ford Motor Company was that he could never get used to working through engineering again—not after the working relationship he and Edsel had developed over the years. How could anyone go from working directly with the president to working for a manager who was three times removed? "I mean," says Gregorie, "I had such a happy relationship with Edsel Ford and working the way I did, directly with him. He was the best man in the business—we were a good fit."

When Oswald joined Ford, he completely revamped Gregorie's and Edsel's patron-directed, designer-run design department into the image of General Motors. In 1953, as part of the modernization program at Ford Motor Company, Oswald moved the design department into a modern new complex across the street from the engineering lab and gave it a new name of the "styling center." It had twelve studios, one for each car and truck line, and for advanced designs. "Certain people ran *this* studio, certain people ran *that* studio," Gregorie explains with a sneer. "There were hundreds of them. It became a boiling pot of backstabbers and politics. I could never have put up with that! I would have gone nuts!" (Figure 19-2)

Figure 19-1. In the more than fifty years since leaving Ford Motor Company, Gregorie has returned to Dearborn only once. In 1974, twenty-eight years after Gregorie had left the company, the members of the design department invited him to Dearborn for a reunion. They took him on a tour of the styling studio, as it was called then, and they held a banquet in his honor at the Henry Ford Museum & Greenfield Village. Here, E.T. Gregorie (right) poses proudly next to his most famous creation, the 1940 Lincoln Continental, on the grounds of Greenfield Village during his visit. (Note the length of Gregorie's pants legs. As in the old days, he still wore his pants legs high above his shoes.) **Standing beside Gregorie** *is his vivacious wife, Evelyn. The gentleman on the left is Dr. Donald A. Shelley, director of the Edison Institute. (Photo courtesy of Henry Ford Museum & Greenfield Village)*

firm and many commitments, and he was upset with Henry Ford II for not paying him what Breech had promised. The exact amount Walker received for the 1949 Ford project is unknown, but it must have been significantly less than the millions of dollars Breech had promised him.

Figure 19-2. This was the Ford design staff, circa 1950. The design department at Ford grew substantially after Gregorie left the company. It was no longer a "country-store" operation. This photograph shows only a few of the many managers now employed in the design department (left to right): unknown; Robert Maguire, design executive, color & trim; Frank Francis, scheduling manager; Hermann Brunn, trim supervisor, color & trim; Arthur Querfeld, design supervisor, color & trim; Gene Bordinat, design manager, Lincoln-Mercury styling director; Don Beyreis, designer, Lincoln-Mercury interiors; George Martin, plaster/plastic fabrication; Al Kellum, supervisor, Lincoln-Mercury modeling staff; Don DeLaRossa, supervisor, Mercury styling; Willys P. Wagner, supervisor, Ford truck styling; Gordon Buehrig, specialist engineering; Charles Stobar, shop supervisor; John Grebe, supervisor, metal shop; William "Bill" Leverenz, design supervisor, Ford styling; Duncan McRae, design supervisor, Ford styling; and Randall Osmon, color specialist, color & trim. (Photo from the Collections of Henry Ford Museum & Greenfield Village)

Meanwhile, Gregorie was enjoying himself in Florida. In 1962, he designed and had built for himself his last yacht. (Figure 19-3) Christened *Drifter*, a name that befit Gregorie's lifestyle since leaving Ford, it was a beautiful forty-four-foot craft, powered by a Cummins diesel engine. Until only a few years ago, Gregorie and his vivacious second wife, Evelyn, whom he married in 1952, had lived on *Drifter*, whiling away their time traveling up and down the Eastern seaboard. They sailed along the sun-drenched beaches of Florida during winters and spent their summers along the shores of Maryland and the Carolinas. "People wonder why I'm so happy," expounds Gregorie. "Well, it's because I got out of the design business just in the nick of time. The few months that I was at Ford while Breech was there, the time came when I decided to weigh anchor, to get underway, so to speak. Tom Hibbard and the others couldn't understand why I was so happy about walking out of a job like that. Well, I too hated to see a good thing go down the drain, but when the time gets to where it isn't a safe anchorage, why, get the hell out of there. So, that's what I did."

Figure 19-3. *This is the last yacht Gregorie designed and had built for himself. Named* Drifter, *it befit his lifestyle after leaving Ford Motor Company. Shown here on deck, Gregorie and his wife, Evelyn, lived on this boat for many years. (Photo courtesy of E.T. Gregorie)*

Years later, Gregorie and his wife docked their boat at Daytona Beach for race day. "We were walking our dog along the pier," explains Gregorie, "when all of a sudden, Henry Ford II popped out of a phone booth. He recognized me immediately."

"What in hell are you doing here?" quizzed Ford.

"Same thing you are," Gregorie told him.

"Why don't you come over to my boat and have a drink?" asked Henry II.

"Oh, we'd like to," Gregorie replied, "but we have our dog here, and…"

"I'll take care of that," interrupted Ford.

"We walked over to his yacht," recalls Gregorie, "and Mr. Ford ordered his captain to walk our dog while the three of us spent the afternoon drinking cocktails and talking about the old days." That was the last time Gregorie saw Henry Ford II.

Today, the Gregories live in a cozy home in northern Florida. "We wanted to stay out on the water longer," laments the sage designer, "but the waterways became too crowded and the people too rough."

Gregorie says he could have been a rich man had Edsel lived, and the industry would have been richer had Gregorie been able to apply his artistic eye to automobile style throughout the 1950s. Were it not for the intervention of Ernest Breech and his transplanted stylists from General Motors, Gregorie could have remained head of Ford styling until at least the early 1960s if not until the early 1970s. During those years, Detroit designed and produced the worst it could offer. If Edsel and Gregorie had been there, they could have at least moderated the trend of the grotesque and conspicuous. "Edsel would have changed with the times," says Gregorie. "He had a terrific forethought. He was a good businessman." Between the two of them, they could have offered "the clean line" against Harley Earl's tail fins and "the pure form" against Virgil Exner's monoliths.

Looking back over the years, Gregorie reflects, "It's hard to say what I would have done to car design had I worked at Ford for another twenty years. A lot of changes occurred in the industry during those years. I liked the styling done by my old friend Frank Hershey while he was at Ford—the Thunderbird and the '57 Ford line. I would have supported designs like those. Today's cars have little character. They all look like a loaf of rye bread. Most of them look like a couple of scoops of ice cream that have set out in the sun too long. The only identity they have is the adornment of gouges in the side and attached moldings. And the grilles on some of today's cars look too pinched.

"But then the designers today are dealing with an entirely different set of conditions than what I faced sixty years ago," continues Gregorie. "Considering what the designers of today have to work with—where the machinery is pushed way forward, and the driver is pushed way forward—some of them do quite well. Some of the new cars look quite smart and nice—not in a sense that I like them—but it's the best today's designers can do with the proportions they are obliged to work with."

"And proportion is the key to handsome designs," says Gregorie. "That's why those '33 and '34 Fords were so beautiful—they were beautifully balanced! The nice, accepted balance for most designers would be to have the weighty affect of the car over the rear axle rather than on the front axle, so that you'd have more of a dimension from the rear axle to the rear end, than from the front axle to the front end. That was considered good design, and that is what we did on the '33 and '34 Fords. On those cars, everything was moved back from the front axle center line, which gave them a light, airy look. Traditionally, having the radiator sitting right above the front axle was considered the proper place to have it located. All those old Packards and Pierce-Arrows and Cadillacs, and who knows what else, had the radiator centered over the front axle. But when the engine compartments were lengthened and new suspension systems developed, the radiator was shoved forward and the grille shoved forward further still. That's when the front end began to over reach the front axle, and we lost the classic proportions of the traditional designs. And before you knew it, we wound up with about two feet of overhang of the front axle! That created a whole new design problem.

"But there is no comparison at all between today's designs and the designs I worked on," concludes Gregorie. "It was a different class of design back in those days. It took different thinking back then, just as it takes different thinking today." (Figure 19.4) For example, when asked what he thinks about the 1995 Continental Mark VIII from Ford, Gregorie quotes Edsel's favorite saying when he was forced to criticize a design: "It looks like they tried too hard." However, who could create a design that compares with Gregorie's Continental—"the most beautiful car in the world"?

Gregorie did not partake in the great transformation that occurred in Ford Motor Company during the 1950s and 1960s, when design became all important and the salary of its chief designer

Figure 19-4. E.T. Gregorie and his wife, Evelyn, in 1986, standing next to one of his creations—a 1948 Lincoln Continental cabriolet. (Photo courtesy of E.T. Gregorie)

went through the roof. However, Gregorie takes solace in the fact that many of his Trade School "boys" survived the purges. Many played significant roles in the reconstruction of the company, while others succeeded in other endeavors.

John Najjar's illustrious career at Ford spanned more than four decades. From his humble beginnings as a design apprentice under Gregorie, Najjar went on to head the Lincoln studio, the truck studio, and the industrial design office at Ford. Some of Najjar's most renowned designs were the 1956 Mercury Turnpike Cruiser, the 1961 Lincoln Continental, the 1964 Mustang (which he named and helped design), and the 1967 Ford pickup, a design that continued with only minor modifications for almost fifteen years. While head of the industrial design office at Ford, Najjar directed the design of the Fairlane Town Center shopping mall and office park, built on property that was once part of Henry Ford's Fair Lane estate. Najjar retired in 1980, after forty-three years of creative service.

Tucker Madawick left Ford during World War II, but not before leaving his mark on Ford automobiles, Jeeps, and the B-24 bomber. After the war, Madawick helped design the Tucker automobile working for J. Gordon Lippincott, contributed to the 1953 line of Studebakers by working for Raymond Loewy, and culminated his design career as design vice president of RCA.

Another one of Gregorie's "boys" who did well was Duncan McRae. After leaving Ford in 1942, McRae worked for Chrysler, and then for Kaiser-Frazer, Studebaker, and Curtiss-Wright. He ended his career at Ford, retiring in 1974.

Bob Thomas also left Ford before World War II, worked for Hudson, and then worked for General Motors. He returned for another stint at Ford in 1947, and then went to work for Nash in 1950. Returning to Ford in 1952, Thomas was instrumental in designing the 1956 Continental Mark II (the first Continental from Ford since 1948) and the highly successful 1961 Lincoln Continental. His last major contribution before retiring in 1974 was the Ford Pinto.

Bud Adams realized early that his heart was not in the body design of automobiles but rather in their mechanics. During the few years he worked for Gregorie in the design department, he took university classes at night and in 1942 graduated with honors from the Lawrence Institute of Technology. After two years of doing military engine design work at Briggs Manufacturing, Adams joined the Navy. Following World War II, he rejoined Ford in its product engineering department. Adams retired from Ford in 1981.

Placid "Benny" Barbera is the only one of Gregorie's original design group remaining with Ford. Amazingly, his career has spanned six decades. During that time, he witnessed the entire evolution of the design department of Ford Motor Company, from the old days until the modern era.

"I'll tell you, when I came here in 1938, we didn't have a dozen people," Barbera recalls, "but we did a lot of work, and we got things accomplished! Mr. Edsel Ford would come in every day after lunch and get together with Mr. Gregorie, and he'd critique the model, and we'd make changes, and in less than three months we had the car all ready to go to engineering. All that and never an hour of overtime! Nowadays, we have close to a thousand people, and we still can't seem to make a schedule.

"There will never be another man like Edsel Ford," Barbera continues. "He was a nice man, a creative man, a visionary. He was a genius as far as design was concerned. The direction came from the top man, Mr. Edsel Ford, and he should get credit for that. And Mr. Gregorie was a master designer. (Figure 19-5) He got his direction from Edsel Ford, and then he followed through and developed the designs. He was a good man to work for, and he got things done. I wish we had Mr. Gregorie and Mr. Ford back!"

Figure 19-5. Taken in November 1945, this is the last official photograph of E.T. Gregorie at Ford Motor Company, and it best exemplifies the great designer. Always confident in his abilities and dapper in his appearance, Gregorie was a light in the stoic Ford Motor Company during the years he worked there. (Photo courtesy of Ford Motor Company)

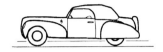

BIBLIOGRAPHY

Books

Adams, G. Eugene "Bud," *40 Years in Product Design and Engineering at Ford Motor Company*, Adams Publishing Co., Green Valley, AZ, 1992.

Armi, Edson C., *The Art of American Car Design*, The Pennsylvania State University Press, University Park, PA, 1988.

Arnold, Marvin, *Lincoln and Continental, Classic Motorcars: The Early Years*, Taylor Publishing Company, Dallas, TX, 1989.

Bonsall, Thomas E., *The Coachbuilt Lincoln*, Turning Point Press, Baltimore, MD, 1995.

Bonsall, Thomas E., *The Lincoln Motorcar: The Complete History of an American Classic*, Stony Run Press, Baltimore, MD, 1992.

Bryan, Ford R., *Henry's Lieutenants*, Wayne State University Press, Detroit, MI, 1993.

Dammann, George H., *Illustrated History of Ford*, Crestline Publishing Co., Sarasota, FL, 1971.

Detroit Style: Automotive Form 1925–1950, Julia P. Henshaw and A.D. Miller, eds., The Detroit Institute of Arts, Detroit, MI, 1985.

Eight Automobiles, The Museum of Modern Art, New York, NY, 1951.

The Ford Book of Styling: A History and Interpretation of Automobile Design, Public Relations, Styling Office, Ford Motor Company, Dearborn, MI, 1963.

General Motors: The First 75 Years, by the editors of *Automobile Quarterly Magazine*, Automobile Quarterly, Inc., Princeton, NJ, 1983.

Hickerson, J. Mel, *Ernie Breech: The Story of His Remarkable Career at General Motors, Ford, and TWA*, Meredith Press, New York, NY, 1968.

Lamm, Michael, and Dave Holls, *A Century of Automotive Style: 100 Years of American Car Design*, Lamm-Morada Publishing Company, Stockton, CA, 1996.

Nevins, Allan, and Frank Ernest Hill, *Ford: Expansion and Challenge 1915–1933*, Charles Scribner's Sons, New York, NY, 1957.

Nevins, Allan, and Frank Ernest Hill, *Ford: Decline and Rebirth 1933–1962*, Charles Scribner's Sons, New York, NY, 1962.

Pfau, Hugo, *The Custom Body Era*, Castle Books, New York, NY, 1970.

Pripps, Robert N., and Andrew Morland, *Ford and Fordson Tractors*, Motorbooks International, Osceola, WI, 1995.

Ritch, Ocee, *The Lincoln Continental*, Motorbooks International, Osceola, WI, 1974.

Sorensen, Charles, *My Forty Years with Ford*, W.W. Norton & Co., Inc., New York, NY, 1956.

Sorensen, Lorin, *The Classy Ford V8*, Silverado Publishing Company, St. Helena, CA, 1982.

Stout, Richard H., *Make 'em Shout Hooray!*, Vantage Press, Inc., New York, NY, 1988.

Thomas, Bob, *Confessions of an Automobile Stylist*, self-published, 1984.

Wilkins, Mira, and Frank Ernest Hill, *American Business Abroad: Ford on Six Continents*, Wayne State University Press, Detroit, MI, 1964.

Periodicals

Borgeson, Griffith, "Ford for Over There: The Perennial Y & C," *Automobile Quarterly*, Vol. 25, No. 2, 1987, pp. 128–147.

Bryan, Ford, and Henry Dominguez, "Remembering Edsel Ford I on the Centennial of His Birth," *The Dearborn Historian*, Vol. 33, No. 4, Autumn 1993.

Dietrich, Raymond H., "Custom Designs Off the Assembly Line," *Wards Quarterly*, Summer 1965.

"Fantastic Ford Finds," *Special Interest Autos*, December 1970, pp. 8–13.

Ford, Edsel B., "The Motor Car of the Future—Manufacturer," *Commercial Art and Industry*, Vol. 19, November 1935.

Foster, Kit, "Edsel's English Enigma: How the Ford 'Special Sports' Became a Jensen," *Automobile Quarterly*, Vol. 36, No. 2, Feb. 1997, pp. 16–31.

"Gentle Gentleman," *Wards Quarterly*, Summer 1965.

Harrison, Bob, "Car Life Classic: 1941 Lincoln Continental," *Car Life*, April 1967, pp. 35–42.

Klucha, Carolyn, "Stroke of Genius," *Heritage Magazine*, October 1988, pp. 27–31.

Lamm, Michael, and David L. Lewis, "The First Mercury and How It Came to Be," *Special Interest Autos*, July–August 1974, pp. 14–53.

Lass, William, "The Meaning of a Name," *Your Edsel Marketer*, Vol. 1, No. 2, publication of Ford Motor Company.

"Personality Profile: John Najjar: Life with 'Mother Ford,' " *Collectible Automobile*, April 1995, pp. 70–77.

Sorensen, Lorin, "The Legend of Gable's Jensen," *White Lady*, Issue 54, a publication of the Association of Jensen Owners, 1990.

Stout, Richard H., "OL5: The Proposed Lincoln for 1940," *The Classic Car*, June 1992, pp. 8–16.

Teague, Walter Dorwin, "Edsel Ford—Designer," *Lincoln-Mercury Times*, Vol. V, May–June 1953, a publication of Ford Motor Company.

Tjaarda, John, "How the Lincoln-Zephyr Was Born," *Motor Trend*, February 1954.

Washburn, Robert Collyer, "Peach Pits, Plaster Cupids, Gold Trim: Many Roads Lead to Motor Car Designing," *Ford Times*, April 1946, pp. 46–51.

Williams, Warren, "The Lincoln Continental: The Dream Becomes Reality," *The Continental Star*, a publication of the Philadelphia Region of the Lincoln Continental Owners' Club, January and April 1988.

Zannis, Vic, "The Last Model A," *Antique Automobile*, September–October 1991, pp. 18–19.

Ford Archives Materials

Adams, G. Eugene "Bud," Design Oral History Project, Edsel B. Ford Design History Center, Henry Ford Museum & Greenfield Village, Dearborn, MI.

Farkus, Eugene, Reminiscences, Accession 65, Henry Ford Museum & Greenfield Village Archives, Dearborn, MI.

Ford, William Clay, Design Oral History Project, Edsel B. Ford Design History Center, Henry Ford Museum & Greenfield Village, Dearborn, MI.

Galamb, Joseph, Reminiscences, Accession 65, Henry Ford Museum & Greenfield Village Archives, Dearborn, MI.

Gregorie, Eugene Turenne "Bob," Design Oral History Project, Edsel B. Ford Design History Center, Henry Ford Museum & Greenfield Village, Dearborn, MI.

Koto, Holden, Design Oral History Project, Edsel B. Ford Design History Center, Henry Ford Museum & Greenfield Village, Dearborn, MI.

Madawick, Tucker P., Design Oral History Project, Edsel B. Ford Design History Center, Henry Ford Museum & Greenfield Village, Dearborn, MI.

Martin, Edward, "Birth of the Continental and Other Observations," unpublished manuscript from the Collections of Henry Ford Museum & Greenfield Village, Dearborn, MI.

Miller, Rhys D., Design Oral History Project, Edsel B. Ford Design History Center, Henry Ford Museum & Greenfield Village, Dearborn, MI.

Najjar, John, Design Oral History Project, Edsel B. Ford Design History Center, Henry Ford Museum & Greenfield Village, Dearborn, MI.

Sorensen, Charles, Reminiscences, Accession 65, Henry Ford Museum & Greenfield Village Archives, Dearborn, MI.

Walker, George W., Design Oral History Project, Edsel B. Ford Design History Center, Henry Ford Museum & Greenfield Village, Dearborn, MI.

Interviews

Barbera, Placid "Benny," various interviews, 1996–1998.

Booth, Wayne, interview, 1996.

Cousins, Ross, various interviews, 1993–1996.

Gregorie, Eugene Turenne "Bob," various interviews, 1991–1998.

Hay, John, interview, July 1996.

Koto, Mrs. Holden, interview, 1996.

Madawick, Tucker P., various interviews, 1995–1996.

Miller, Rhys D., interview, 1996.

Najjar, John, various interviews, 1994–1998.

Olinik, Andrew, interview, 1996.

O'Rear, Emmett, various interviews, 1995–1998.

Roberts, Ralph, various interviews, 1989–1996.

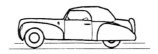

INDEX

Adams, Eugene "Bud," 94
    career, 316
    clay modeling, 101
    design of 1939 Mercury, 220
    design of 1941 Ford, 225
    design of Continental, 156-158
    design department role, 131, 169
    photo, 96
    relationship with E.T. Gregorie, 118, 125
Allan, Letha, 118, 122
Astaire, Fred, 51
Australian Job automobile, 36-37, 38
Avery, Clarence, 25

Bacon, Irving, 28, 107, 110
Baird, Harvey, 121
Barbera, Placid "Benny," 2, 7, 97, 104, 107, 125, 141, 169
    career, 316
    photo, 96, 120
Baxter, Richard, 122
Bendix Aviation Corporation, 285
Beneicke, Dick, 98, 100, 101, 104, 120, 125, 149, 169, 205
Bennett, Harry, 27, 279
Bentley automobile, Museum of Modern Art auto design critique, 80
Beyreis, Don, 312

"Big K Lincoln," 155, 166
Bliss, Raymond, 120
Bogre, Michael, 102, 123
Boice, Carrol, 124
Booth, Wayne, 101, 103, 121
Bordinat, Gene, 312
Breech, Ernest
    Gregorie and Walker design competition, 291, 293, 295, 306
    joins Ford, 285-287
    naming Edsel car, 21
    relationship with E.T. Gregorie, 9, 242, 306, 309-311, 314
    relationship with George Walker, 306
Breer, Carl, 7
Breitmayer, Vern, 122, 303
Brewster & Company, 14
Briggs Manufacturing
    1932 Ford, 54-56
    1935 Ford, 56
    1936 Ford, 56
    1937 Ford, 45, 145-147
    designers of, 8, 14, 56, 58, 98, 140, 142, 178, 179
    Model Y, 60
    Sterkenberg design, 75
Briggs, Walter, 53
Brown, George, 37-38

Brunn & Company, 40, 116
Brunn, Hermann, 42, 116, 122, 312
Buehrig, Gordon, 312
Bugas, John, 242

Cadillac automobile, Museum of Modern Art auto design critique, 80
Campbell, Elre, 118, 122
Campsall, Frank, 280
Carroll, Leota, 118, 119, 121
Cartier, Pierre, 48
Caspers, Francis, 122
Cheek, John, 123
Chevrolet automobile, 36, 43, 44
Chrysler, Airflow design, 75
Citroën automobile, 1925 model, 16-18
Clark, Lawrence, 121
Clark, William, 124
Clay modeling, 8, 95, 98, 100, 101-105, 129
Comstock, Gov. William, 68
Continental automobile
    1940 model, 164, 165
    1942 model, 237, 239, 241
    1946 model, 241
    early design, 155-156, 158
"Continental cars," 8, 67-69, 71-73, 81-87, 168
Cord automobile, Museum of Modern Art auto design critique, 80
Cousins, Ross, 2, 8, 86, 276
    career, 112
    design of Continental, 161
    drawings by, 99, 251
    photos, 113, 118, 121
Cox & Stevens, 13, 18
Crawford, John, 220
Crecelius, Henry, 19, 70, 104
Crusoe, Lewis, 286

D'Angelo, Romeo, 121
Davis, John, 136, 201, 247, 283, 308
DeLaRossa, Don, 312
Dickason, Doris, 118, 122
Dietrich, Ray, 14, 15, 42, 48, 51, 132
Dobben, John, 116, 123

Doehler, Robert, 123
Dolan, Robert, 121
Doss, Henry, 136
Doyle, James, 120
Duncan, John, 102, 124

Earl, Harley, 2, 3, 4, 9, 14, 18, 54, 153, 195, 199
Ebling, George, 162
Edsel automobile, 21-22
Ehlendt, Peter, 123
Elco Works, 12, 13
Engel, Elwood, 291
England, Gerald, 122
English Finance Act, 57
Esper, Al, 253
Evans, Melvin, 124
Exner, Virgil, 9

Falcon automobile, 307
Farkus, Gene
    blackboard drawings, 37, 43
    Model Y, 58
    photo, 224
    small V-8 engine, 248-249
Ferguson, Harry, 193
Ferguson, Robert, 124
Flajole, Bill, 54
Fleming, William, 21-22
Ford, Benson, 22
Ford, Edsel
    appearance, 22, 23
    artistic abilities of, 28-30
    automobile design
        Australian Job, 36-37, 38
        1939 Continental, 155-168
        1940 Continental, 164, 165
        1942 Continental, 237, 239, 241
        1946 Continental, 241
        "continental cars," 8, 67-69, 71-73, 81-87, 168
        1937 Ford, 142, 144-148
        1938 Ford, 144, 148, 149, 169, 171, 172-180, 195, 199
        1939 Ford, 180, 195, 198, 199

Ford, Edsel *(continued)*
   automobile design *(continued)*
      1940 Ford, 195, 196, 198, 199
      1941 Ford, 213-223, 225
      1942 Ford, 213, 227-233
      1946 Ford, 235, 238
      Ford V-8, 53-56, 63-66, 140-144, 169, 171, 174-176, 180, 183-186, 195
      Lincoln-Zephyr, 8, 75-80, 149-153, 155-156, 158, 165, 187, 188
      1939 Mercury, 201-211
      1941 Mercury, 225-227
      1942 Mercury, 227, 234
      1946 Mercury, 235, 236
      Model T, 35-38, 43-44, 46, 52
      Model Y, 57-62
      speedsters, 31, 33-35
      Torpedo roadster, 35
   autos owned by, 31-34
      "continental car," 72, 278
      1941 Fords, 278
      1941 Lincoln Continental, 278
      1923 Lincoln coupe, 40
      1941 Mercury, 278
      Model A, 31
      Model N, 31, 32
      Model T, 31, 33
      1912 Rolls-Royce Silver Ghost touring car, 41
   boats owned by, 51-53
   color palettes, 44, 48, 51
   competing against Chevrolet, 36, 43, 44
   custom-body companies, 40-42
   death, 9, 73, 242, 276
   design department, establishment of, 14, 19, 70, 91-93
   design style, 2-3, 6, 51, 63, 133
   early life, 22
   early professional life, 22, 24
   favorite color, 51
   Forditis, 272
   good taste, 51-52
   Guardian Group, 68
   homes of
      Gaukler Pointe, 28, 72, 276
      Grosse Pointe, 73, 125
      Hobe Sound, 91, 272, 274
   illness, 262, 272-273, 276
   Lincoln automobiles, 38
   management style, 24-27
   personality, 22
   photos, 23, 25, 32-34, 39, 49, 50, 52, 65, 90, 135, 277
   relationship with father, 22, 24, 26-28, 135, 136, 202
   relationship with E.T. Gregorie, 3, 20, 127, 132-134, 137, 201-202
   Round Table meetings, 135-136
   sports car, 71, 87, 89, 168
Ford, Eleanor, 22, 28, 39, 52
Ford, Henry
   business decisions, 27
   chassis design, 7, 35, 63, 66, 81, 196
   "continental car," 86, 87
   design constraints on Gregorie, 6-7
   design style, 7, 35
   favorite color, 44, 48
   management style, 24-27
   meeting with Gregorie, 278-279
   photos, 25, 49, 65, 135
   relationship with design department, 14
   relationship with son, 22, 24, 26-28, 135, 136, 202, 295
   Round Table meetings, 135-136
   smoking and, 98
Ford, Henry, II, 22, 279, 283-287
   construction program, 308
   Falcon automobile, 307
   photo, 308
   policy committee, 283, 286
   relationship with E.T. Gregorie, 9, 224, 279-280, 283, 313
   Sportsman, 242-245
Ford Motor Company
   aircraft factory, 67, 68-70, 71, 242
   automobiles. *See* Ford vehicles
   chassis, 6-7, 66, 81, 139, 196
   commercial line, 129, 190-194
   competing against Chevrolet, 36, 43, 44
   Dearborn Inn, 68, 70
   Depression and, 67-68

Ford Motor Company *(continued)*
- design department, 3, 91-137
  - appearance, 96-98, 99
  - body layout and engineering, 104
  - clay modeling, 8, 95, 98, 100, 101-105, 129
  - design process, 129-131
  - establishment of, 14, 19, 70, 91-93
  - fabrication shop, 104
  - French curves, 131-132
  - male designers only, 98
  - management of, 4-5
  - photos, 96, 97, 99, 113, 114, 128, 312
  - staff, 104-127, 311, 312
  - women in, 98, 116, 118
- Engine & Electrical Engineering Building, 93-94, 98
- Ford Apprentice School, 107
- General Motors, competition with, 75
- Henry Ford Trade School, 107-109
- postwar financial condition, 284-285
- postwar vehicles, 235, 284, 287, 288, 290
- Rouge plant, 101
- smoking on premises, 98
- transformation in 1950s and 1960s, 314-316
- Tri-Motor airplanes, 67, 70, 104

Ford V-8, 8
- 1933-1934 models, 63-66, 314
- 1935 model, 140-142
- 1936 model, 140, 142-144
- Deluxe, 75, 113, 169, 171, 174, 176, 180, 183-186, 195, 211
- design of, 53-56
- Standard, 169, 171, 174, 175, 180, 195, 211
- Super Deluxe, 227

Ford vehicles
- Australian Job, 36-37, 38
- chassis, 6-7, 63, 66, 81, 139, 196
- commercial line, 273, 284
- 1939 Continental, 155-168
- 1940 Continental, 164, 165
- 1942 Continental, 237, 239, 241
- 1946 Continental, 241
- "continental cars," 8, 67-69, 71-73, 81-87, 168
- design process, 129-131

Edsel, 21-22
Falcon, 307
five-cylinder engine, 249
1937 Ford, 142, 144-148
1938 Ford, 144, 148, 149, 169, 171, 172-180, 195, 199
1939 Ford, 180, 195, 198, 199
1940 Ford, 195, 196, 198, 199
1941 Ford, 213-223, 225
1942 Ford, 213, 227-233
1946 Ford, 235, 238
1949 Ford, 293-298
1939 Ford coupe, 170
Ford V-8, 8
- 1933-1934 models, 63-66, 314
- 1935 model, 140-142
- 1936 model, 140, 142-144
- Deluxe, 75, 113, 169, 171, 174, 176, 180, 183-186, 195, 211
- design of, 53-56
- Standard, 169, 171, 174, 175, 180, 195, 211

front-wheel drive, 253, 254
Lincoln
- 1939 model, 211
- 1942 model, 236
- design of, 19-20, 40-42
- fastback design, 300
- postwar model, 272, 295, 299, 305

Lincoln Continental
- 1940 model, 311
- 1948 model, 315
- 1961 model, 110
- postwar model, 252

Lincoln-Zephyr, 165
- 1936 model, 8, 75-80
- 1938 model, 149-153
- 1939 model, 8, 160, 211
- 1940 model, 187, 188
- Continental design from, 155-156, 158

Mercury, 75
- 1939 model, 201-211
- 1941 model, 225-227
- 1942 model, 227, 234
- 1946 model, 235, 236

Ford vehicles *(continued)*
    Mercury *(continued)*
        1949 model, 8, 306
        postwar model, 187, 295, 303
        Tudor sedan, 289
        Turnpike Cruiser, 110
    Model A, 3, 8, 196, 197
        colors, 48
        design changes, 53
        design of, 44-48, 50
        owned by Edsel Ford, 31
    Model N, 3, 32
    Model T, 3, 16, 27
        Australian Job, 36-37, 38
        design changes, 35-38, 43-44, 46, 52
        owned by Edsel Ford, 31, 33
        photo, 46
        speedster, 31, 33-35
        Torpedo roadster, 35
    Model Y, 57-62
    Mustang, 110
    Pinto, 316
    postwar models, 247-273
    Sportsman, 242-245
    Torpedo, 35
    touring cars, 20
    V-8 engines, 248-249
    Vedette, 291, 293, 294
    wooden-bodied vehicles, 242
    X-8 engine, 249
Ford, William, 21-22, 161
Ford Apprentice School, 107
Fox, William, 120
Framke, Werner, 120, 302
Francis, Francis T. "Frank," 113, 121, 125, 312
French curves, design tools, 131-132
Fulton, Gordon, 120, 124

Gable, Clark, 87
Galamb, Joseph, 20, 247
    career, 3
    design of Model A, 44, 47
    design of Model T, 36, 42, 44
    firing of, 224, 284
    photo, 224
    relationship with Henry Ford, 5, 6, 280
Galla, Stephen, 123
Ganzenhuber, Emil, 123
General Motors Company, 52
    Art & Colour Section, 14, 15
    Bendix division, 285
    Chevrolet, 36, 43, 44
    competition with Ford, 75
    grille designs, 153
    hired Gregorie, 15
    vertical-mounted engine, 253
Graham, Dr. Roscoe, 273
Grebe, John, 312
Gregorie, Eugene T.
    appearance, 2, 11, 12
    automobile accident, 17-18
    autos designed by, 3, 8-9
        1939 Continental, 155-168
        1942 Continental, 237, 239, 241
        1946 Continental, 241
        "continental cars," 8, 67-69, 71-73, 81-87, 168
        1938 Ford, 144, 148, 149, 169, 171, 172-180, 195, 199
        1939 Ford, 180, 195, 198, 199
        1940 Ford, 195, 196, 198, 199
        1941 Ford, 213-223, 225
        1942 Ford, 213, 227-233
        1946 Ford, 235, 238
        1949 Ford, 293-298
        Ford V-8, 8, 53-56, 63-66, 140-144, 174-176, 180, 183-186, 195
        front-wheel drive, 253, 254
        Lincoln-Zephyr, 8, 75-80, 149-153, 155-156, 158, 165, 187, 188
        1939 Mercury, 201-211
        1941 Mercury, 225-227
        1942 Mercury, 227, 234
        1946 Mercury, 235, 236
        1949 Mercury, 8
        Sportsman, 242-245

Gregorie, Eugene T. *(continued)*
   autos owned by, 15
      1925 Citroën, 16-18
      "continental car," 86
      1928 Eifel, 61
      1929 La Salle coupe, 17
      1947 Lincoln Continental, 241
      Model A beach wagon, 242
      Model Y special touring, 62
      Model Y Taunus, 62
      Model Y Tudor, 62
      1930 Stutz roadster, 125, 126
      1930 Whippet roadster, 18
   begins work at Ford, 1-2
   birth, 11
   boating and, 2, 4, 125, 127, 313
   Brewster & Company, 14
   chassis design, 66, 81, 139, 143, 196
   childhood, 11-12
   clay models, 8, 95, 98, 100, 101-105, 129
   Cox & Stevens, 13, 18
   death of Edsel Ford, 280
   design constraints on, 6-7
   design department, 91-137
      establishment of, 14, 19, 70, 91-93
      management and, 4-5
      management style, 125, 133
      office area, 98
   design style, 2, 3, 35, 66, 133, 143, 314
   Elco Works, 12, 13
   England trip, 68
   fired after Edsel's death, 280
   General Motors employment, 15-16, 18
   laid off, 68
   lifestyle, 3-4
   marine architect, 12-14, 18
   meeting with Henry Ford, 278-279
   miniature railroad, 4
   own design firm, 281, 284
   personality, 2
   photos, 12, 96, 128, 214, 224, 241, 285, 307, 308, 311, 317
   post-Edsel Ford period, 286-287, 291, 293, 295, 299, 306-307
   quits Ford Motor Co., 9, 310
   relationship with designers, 2, 6
   relationship with Edsel Ford, 3, 20, 127, 132-134, 137
   relationship with Joe Galamb, 5, 280
   relationship with Henry Ford II, 9, 224, 279-280, 281, 283, 283-287, 313
   relationship with Henry Ford staff, 5-6
   relationship with Larry Sheldrick, 5
   retirement, 310, 313
Gregorie, Eugene T., Sr., 11, 13
Gregorie, Evelyn, 311, 313, 315
Guardian Group, 68
Gundy, Elmer, 121

Harder, Delmar, 286
Haskins, Jay, 123
Haven, Kenneth, 14-15
Hay, John, 4, 112, 115, 124
Henry Ford Trade School, 107-109
Hershey, Franklin Quick, 15-16, 314
H.H. Franklin Manufacturing Company, 14
Hibbard, Thomas, 14
   design staff role, 122, 283, 284, 310
   LeBaron, 116
   photo, 102, 122
   postwar Ford design, 291
   quits Ford, 311
Hobbs, T.C., 122
Hoffman, Dutch, 124
Homanick, Michael, 120
Hooper & Company, 68
Huggins, James, 122, 306
Hughson, Billy, 199

Iafrate, Arseno, 120
Inskip, John, 14

J.B. Judkins, 40
Jensen, Allan, 88, 90
Jensen automobile, 88-89, 168
Jensen Motors, Ltd., 88-89, 90

Jensen, Richard, 88, 90
Johnson, Frank, 149, 224
Judkins Company, 116

Kahn, Harold, 88
Kanzler, Ernest, 134
Karpeles, Arthur, 120, 301
Karstadt, Clarence, 54
Keck, Walter, 121
Kellum, Al, 304, 312
Klingensmith, Frank, 26
Knudsen, William, 26
Kolt, Bruno, 104, 107, 113, 121, 169
Koto, Holden, 54, 56, 140, 291
Kramer, Clare, 63-64, 131
Kroll, Ben, 121
Kruke, Walter, 84, 85, 97, 104, 124, 249

La Salle coupe, 17
LaBreck, Glen, 120
Lang, Victor, 23
LeBaron, Carrossiers, 40, 53, 116
Leland, Henry, 38, 43
Leland, Wilfred, 38, 43
Lepine, Albert, 273
Leverenz, William M., 98, 104, 120, 312
Levy, James, 123
Lewis Hotel (Detroit), 15
Lincoln automobile, 27
    1939 model, 211
    1942 model, 236
    design of, 19-20
    fastback design, 300
    Model K, 155, 156, 165, 166
    postwar model, 272, 292, 299, 305
Lincoln Continental automobile, 110, 159, 252, 311, 315
Lincoln Motor Company, 18, 38
Lincoln-Zephyr automobile, 165
    1936 model, 8, 75-80
    1938 model, 149-153
    1939 model, 160, 211
    1940 model, 187, 188
    Continental design from, 155-156, 158
    Museum of Modern Art auto design critique, 80
Lippincott, J. Gordon, 116, 316
Locke & Co., 40, 116
Loewy, Raymond, 3, 116, 316
Lucht, Harold, 120
Lynch, James, 104, 112, 124, 303

McCarroll, Russell, 285, 308
McKee, Joseph, 124
McKenzie, John, 124
McLean, Ross, 120
McMaster, F., 122
McRae, Duncan, 117, 312, 316
Madawick, Tucker, 4, 116, 125, 195, 210, 247, 315
Maguire, Robert, 312
Martin, Edward, 99, 104, 111
Martin, George, 312
Martin, Pete, 5, 135, 155, 169, 220
    "continental car," 86, 87
    photo, 224
Mayo, William, 26-27
M.B.K. Motors, 88
Mearns, James "Jimmy," 94, 95, 102, 105, 124
Mecklenburg, Ray, 102, 120
Mercedes automobile, 13-14
Mercury automobile, 75
    1939 model, 189, 201-211
    1941 model, 225-227
    1949 model, 8, 306
    postwar model, 187, 295, 303
    Tudor sedan, 289
    Turnpike Cruiser, 110
Miller, Charles, 120, 121
Miller, Rhys, 54, 140
Mitchell, Fred, 106
Model A Ford, 3, 8, 196, 197
    beach wagon, 242-245
    colors, 48
    design changes, 53

Model A Ford *(continued)*
    design of, 44-48, 50
    owned by Edsel Ford, 31
    photo, 49
Model K Lincoln, 155, 156, 165, 166
Model N Ford, 3, 31, 32
Model T Ford, 3, 16, 27
    Australian Job, 36-37, 38
    design changes, 35-38, 43-44, 52
    owned by Edsel Ford, 31, 33
    photo, 46
    speedster, 31, 33-35
    Torpedo roadster, 35
Model Y Ford, 57-62
    photo, 59
    special touring model, 62
    Taunus, 61-62
    Tudor model, 62
Modeling bridges, 130, 131
Moore-Brabazon, Lt. Col. John, 87-88
Morrison, James, 121
Motor Products Company, 18
Murray Corporation, 44-46, 147
Museum of Modern Art
    1936 Zephyr, comment on, 8
    auto design critiques, 80
    award to Continental, 168
Musial, Walter, 124
Mustang automobile, 110

Najjar, John, 7, 13, 94, 116, 121, 127, 131, 137
    autos designed, 315
    clay modeling, 101
    joins Ford, 107
    late career, 315
    photos, 110, 111, 114, 128
    "soy bean car," 165
Newton, Charley, 110
Noneman, Martin, 124
Northrup, Amos, 3, 44, 148

O'Leary, Howard, 15
O'Rear, Emmett, 94, 96, 97, 99, 107, 109, 113, 124, 126, 169
O'Roke, Beth, 118, 122
Oros, Joe, 291
Osmon, Randall, 124, 312
Oswald, John, 311

Paton, Clyde, 285
Paulson, Robert, 105
Perry, Lord Percival, 57, 61, 62, 68, 87
Peterson, Florence, 118, 121
Pierce-Arrow, 15
Pierrot, George, 51
Pinto vehicle, 316
Pioch, William, 131, 224, 229
Prance, Al, 54, 56
Prefect automobile, 61
Preston, Harry, 124
Pugh, Bert, 104, 123

Querfeld, Arthur, 312

Ramstrum, Eric, 120
Rands, William, 18
Rankin, Robert, 272
Regitko, Martin, 104, 112, 113, 116, 123, 131, 157, 158, 254
Regitko, Ted, 123
Roberts, Ralph, 14, 53, 56, 75, 142
    1935 and 1936 models, 140
    Briggs Manufacturing, 53-54, 75, 142
    LeBaron, Carrossiers, 53
    Model Y design, 58, 60
    photo, 56
Roeder, Dale, 224, 285
Rolls-Royce Silver Ghost touring car, 41
Rootes Body Company, 68
Roper, Elmo, 291
Rosenau, Fred, 123
Ruddiman, Catherine, 28, 30
Rusk, John, 124

Schmidt, William, 122
Schuch, Anton, 105, 124
Scott, Ed, 224
Sheldrick, Larry, 5, 57, 84, 139, 195, 220, 224, 253, 286
Shelley, Dr. Donald A., 311
Shumann, Roy, 110
Skelton, Owen, 7
Skinner, Richard, 121
Sloan, Alfred, 52, 195
Sommer, Herman, 120
Somogyi, Joseph, 102, 124
Sorensen, Charles, 5, 24, 38, 44, 48, 220
    1937 Ford, 144
    "continental car," 86, 87
    firing of, 224
    interferes with design, 219, 222
    photo, 135, 224
    relationship with E.T. Gregorie, 127
    Round Table meetings, 135, 136
"Soy bean car," 165
Sportsman automobile, 242-245
Standard Motor Company, 88
Stephenson, Tom, 131
Sterkenberg design, 75-77
Stevenson, Charles, 120
Stobar, Charles, 312
Studebaker automobile, 291, 292
Stutz roadster, 125, 126
Summerville, Clifford, 121

Tallberg, Val, 161, 285
Tarran, Robert, 121
Tasman, George, 116, 120
Taunus automobile, 61-62
Testaguzza, Cesare, 114, 120, 305
Testaguzza, Gino, 305
Thomas, Robert McGuffey, 96, 97, 107, 110, 113, 161, 316
Tjaarda, John, 4
    Briggs Manufacturing, 3, 53-54, 56, 140
    photo, 56
    Sterkenberg design, 75-77
Todd, Robert, 123
Touring cars, 20
Tremulis, Alex, 54
Tri-Motor airplanes, 67, 70, 104
Trombley, Clyde, 120, 130, 302
Tucker automobile, 116

Upton, Clyde, 124

V-8 engines, 248-249
Vedette automobile, 291, 293, 294
Velten, Richard, 123

Wagner, Willys, 97, 104, 106, 107, 121, 125, 169, 247, 312
Walker, George W., 286-287, 293, 306, 309
Wallis, George, 121
Walter, John, 96, 97, 104, 107, 281
Waterhouse, Charlie, 116
Welling, Rudie, 123
Wengren, Walt, 54
Wharam, Jack, 224, 285
Whippet roadster, 18
White Lady automobile, 90
Wibel, Al, 135, 220
Willoughby & Company, 40
Wilson, Jim, 54, 56
Wojtas, Anthony, 124
Wright, Frank Lloyd, 8, 168
Wright, Phil, 54, 56, 140, 142
W.S. Smith & Sons, 88

Youngren, Harold T., 286, 291, 309, 310

Zawacki, Francis, 107, 125
Zeder, Fred, 7
Zephyr automobile. *See* Lincoln-Zephyr automobile
Zoerlein, Emil, 136

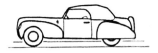

ABOUT THE AUTHOR

Henry L. Dominguez has studied Ford history for more than twenty years. As a teenager, he became interested in Ford automobiles and purchased his first car—a 1946 Ford business coupe—for only $59.00 at an estate sale. Since then, he has owned many old Fords: a 1927 Model T coupe, several 1940 and 1946–1948 Fords, and a 1951 Ford woodie wagon. His present old Ford is a 1940 Ford Deluxe coupe. In addition to restoring these vintage Fords, Mr. Dominguez developed a strong interest in Ford Motor Company History. In 1980, he published his first book about dealerships within the Ford organization. In writing about Ford, he has interviewed many people who have personally known Henry Ford and Edsel Ford, including E.T. Gregorie.

Mr. Dominguez holds a B.S. from Weber State University (Ogden, Utah) and an M.B.A. from the University of New Mexico (Albuquerque, New Mexico). After college graduation, he began work in 1976 as a design engineer for the Saginaw Steering Gear Division of General Motors. In 1988, he joined the Saturn Corporation and currently is a field service engineer there, covering Northern California and the Northwest.

Mr. Dominguez is a member of the Society of Automotive Engineers, the Society of Automotive Historians, and The Henry Ford Heritage Association.